W9-CJK-172

3 1205 00416 0410

Basic Electronics for Tomorrow's Inventors

PUBLIC LIBRARY
DANVILLE, ILLINOIS

About the Author

Nick Dossis lives in England and holds various qualifications in electronic engineering. He has been playing around with electronics for most of his life, originally encouraged by his grandfather, who bought him his first crystal radio set when he was about seven years old. Nick has also written *Brilliant LED Projects: 20 Electronic Designs for Artists, Hobbyists, and Experimenters*, another electronics book published by McGraw-Hill Professional (2011). He continues to design and build electronic circuits in his spare time.

Basic Electronics for Tomorrow's Inventors

A Thames & Kosmos Book

Nick Dossis

New York Chicago San Francisco
Lisbon London Madrid Mexico City
Milan New Delhi San Juan
Seoul Singapore Sydney Toronto

The McGraw-Hill Companies

Cataloging-in-Publication Data is on file with the Library of Congress

McGraw-Hill books are available at special quantity discounts to use as premiums and sales promotions, or for use in corporate training programs. To contact a representative, please e-mail us at bulksales@mcgraw-hill.com.

Basic Electronics for Tomorrow's Inventors: A Thames & Kosmos Book

Copyright © 2013 by The McGraw-Hill Companies. All rights reserved. Printed in the United States of America. Except as permitted under the Copyright Act of 1976, no part of this publication may be reproduced or distributed in any form or by any means, or stored in a database or retrieval system, without the prior written permission of publisher.

McGraw-Hill, the McGraw-Hill Publishing logo, TAB™, and related trade dress are trademarks or registered trademarks of The McGraw-Hill Companies and/or its affiliates in the United States and other countries and may not be used without written permission. All other trademarks are the property of their respective owners. The McGraw-Hill Companies is not associated with any product or vendor mentioned in this book.

Thames & Kosmos is a registered trademark of Thames & Kosmos, LLC.

1234567890 QDB QDB 1098765432

ISBN   978-0-07-179469-5
MHID   0-07-179469-7

Sponsoring Editor: Roger Stewart

Editorial Supervisor: Janet Walden

Project Manager: Nidhi Chopra, Cenveo Publisher Services

Acquisitions Coordinator: Molly Wyand

Copy Editor: Lisa Theobald

Proofreader: Lisa McCoy

Indexer: Valerie Haynes Perry

Production Supervisor: Jean Bodeaux

Composition: Cenveo Publisher Services

Illustration: Cenveo Publisher Services and Greg Scott

Art Director, Cover: Jeff Weeks

Cover Designer: Jeff Weeks

Series Design: Mary McKeon

Information has been obtained by McGraw-Hill from sources believed to be reliable. However, because of the possibility of human or mechanical error by our sources, McGraw-Hill, or others, McGraw-Hill does not guarantee the accuracy, adequacy, or completeness of any information and is not responsible for any errors or omissions or the results obtained from the use of such information.

This is for Elissa, my soul mate. I couldn't have done this without you.

PUBLIC LIBRARY
DANVILLE, ILLINOIS

Contents

Acknowledgments

Special thanks go to my family for supporting me as I wrote this book, especially Jasmine and Georgia Dossis, who helped me take a lot of the close-up photography, and who also doubled up as models in some of the photographs. I'd also like to thank Roger Stewart of McGraw-Hill Professional for giving me the opportunity to write this book and fulfill another one of my dreams.

Introduction

If you are reading this book, you might be a newcomer to electronics who wants to understand the basics so that you can create some interesting circuits. Or you may have already dabbled in the art of electronics and you want to learn some more. Either way, this book will be useful to anyone who is interested in learning about electronics, and it also aims to be a useful resource for electronic hobbyists of all ages and ability levels. Younger readers might find it useful to have an adult around to help them to get started; however, the circuit diagrams and detailed close-up photographs contained in each chapter make it really easy to follow and build the experiments.

The projects and experiments contained in each chapter use inexpensive, readily available electronic components that you can buy from local electronics stores and many electronic suppliers on the Internet. Also, you don't need to be an expert at soldering to build these experiments, because no soldering is required! All the projects and experiments use breadboard, which creates a "plug-and-play" environment for you to build your electronic circuits.

What's Inside the Book?

The book is split into five parts, and I recommend you read the first part, "Let's Get Started," before you build any of the experiments, because it explains some important concepts that you'll need in order to work through the book. You might also find this part useful as a reference as you read through the book. It introduces the common equipment that you will need, along with the basics about electronic building blocks and the components that you will come across in each experimental part of the book.

The next four parts of the book are packed with experiments and real-world examples that help you understand how some of these devices work. In each part of the book, you will identify some of the electronic building blocks that go into each everyday device discussed.

What Does Each Chapter Contain?

Each experimental chapter starts by providing an introduction to the experiment and then includes the following sections:

- **The Circuit Diagram** The circuit diagram shows how each of the electronic components are connected together to produce the device in each experiment.

- **How the Circuit Works** This section describes the circuit diagram and explains how each part of the circuit works. This section is important to read, because it identifies the building blocks used to make the circuit and also helps you to learn how to read circuit diagrams, which are necessary for creating any type of circuitry.

- **Things You'll Need** This section lists all of the electronic components and equipment that you'll need to build the experiment.

- **The Breadboard Layout** Plenty of close-up photographs of the breadboard layout are included in each chapter, and some of these are also taken at different angles to give you a better perspective. You will use these photographs as a guide to building each electronic circuit.

- **Time to Experiment!** This is the fun part. It shows you how to get your experiment working.

- **Summary** The end of each experiment includes a summary of what you have learned in the chapter, and it also makes some suggestions about other uses for the circuit that you have built.

Experiment Difficulty

Some experiments are more difficult to build than others, so the complexity level of each experiment is indicated by the following symbols, shown next to the experiment heading in each chapter. Three different categories of experiments are included in the book:

 Beginner These experiments are really easy to build and should be easy to follow by any newcomer to electronics. These experiments also outline some important basic electronic principles.

 Apprentice These experiments are slightly more complex to build than the beginner-level experiments. Younger readers may require some parental assistance when following these chapters.

 Inventor These experiments are meant to be tackled by readers who are more experienced in electronics and have already built a number of the beginner- and apprentice-level experiments.

A Note from the Author

I personally started experimenting with electronics when I was around six or seven years old and quickly became hooked. Even though I am over 40 years old now, I still love connecting a handful of components together and bringing them alive. You will soon see how this can be done easily and at very little cost. For me, electronics is an enjoyable and an inexpensive hobby, and I hope that you enjoy reading this book and use the knowledge within as a basis of inventing your own projects in the future. Understanding electronics can be the underpinning of many different careers in engineering, science, and business entrepreneurship. And, who knows, if you get the electronics bug, you could be the next Steve Jobs or Steve Wozniak of the future!

Each project and experiment has been extensively tested as part of writing this book; however, the author cannot guarantee the long-term performance, or accept legal responsibility for the results of building them. The reader builds the projects and experiments outlined in this book at his or her own risk.

PART ONE

Let's Get Started!

CHAPTER 1

Setting Up Your Workbench

BEFORE YOU START TO LEARN ABOUT electronics and electronic components, you will need to put together a few pieces of equipment that are required to build the experiments in this book. Figure 1-1 shows some of the basic tools you'll need. Each piece of equipment is described in more detail in this chapter.

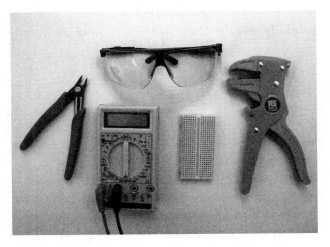

FIGURE 1-1 Some of the equipment you'll need to build the experiments in this book. From top, moving clockwise: safety glasses, wire strippers, breadboard, multimeter, and wire cutters.

INTERESTING FACT

Electronic components are the individual building blocks of every *electronic circuit*. You will be learning about some of the many different types of electronic components in Chapter 3. Each electronic component contains a number of attached wires and these allow you to connect several components together; these wires are sometimes called *component leads*.

Breadboard

Each of the experiments in this book shows you how to build an *electronic circuit*. If you are new to electronics, you might be wondering just what an electronic circuit is. You create an electronic circuit by joining several electronic components together using electrical wire. The results of creating a circuit can be complex or simple, such as causing a light to flash on and off.

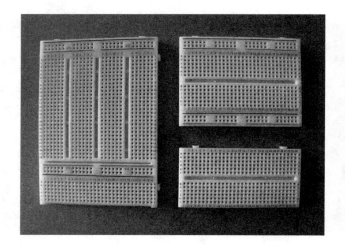

FIGURE 1-2 Various sizes of breadboard

FIGURE 1-3 The gray lines show how the internal connections of a typical breadboard connect various holes together.

You can build electronic circuits in several ways, but one of the easiest ways is to use a *breadboard*. Despite its name, a breadboard isn't something you cut bread on! An electronic breadboard, or plugboard as it sometimes called, is a plastic panel that lets you connect and build electronic circuits without requiring the use of special equipment such as a soldering iron. Figure 1-2 shows some typical sizes and configurations of breadboard; these are the types that I used when building the experiments in this book.

Breadboard comes in many different sizes, shapes, and configurations, so you don't need to use the same type of breadboard that I use. What you do need to understand, however, is how breadboard allows you to connect electronic components together. Notice that a breadboard contains lots of holes, which are sometimes identified by letters and numbers. Hidden inside the breadboard, underneath each of the holes, are connector strips that electronically link various holes together inside the board; you'll use these holes to connect multiple *component leads*—the electrical wires—together.

Figure 1-3 shows an example of how these internal electronic connections are configured inside a typical breadboard.

Each hole in the breadboard is large enough for you to push an electronic component lead into it. The electronic connection inside the board then "grabs" the lead and makes an electrical connection to the other holes in the same row. The gray lines in Figure 1-3 show, for example, that holes a1, b1, c1, d1, e1, and f1 are all connected together. If you want to change a circuit layout, it's simple: just carefully pull the component leads out of the board and start over.

 NOTE

Some breadboard layouts may differ from those that I used to write this book, so always refer to the manufacturer's instructions to identify which holes are connected together. If you use a different type of breadboard, you may need to modify the component layouts slightly to those shown in each chapter.

Each experiment in this book will show you a circuit diagram that will be explained in detail, and close-up photographs will show you how you can build the circuit on a piece of breadboard. Figure 1-4

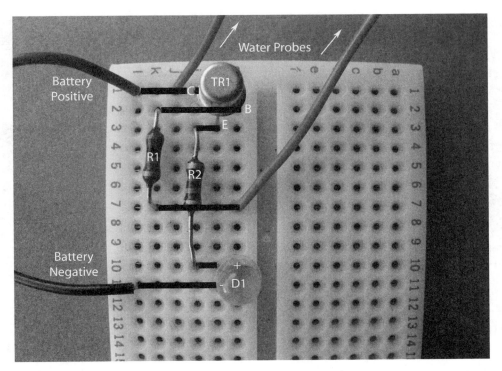

FIGURE 1-4 Each experiment shows a number of close-up photographs to help you build the breadboard layout.

shows you what a breadboard layout looks like (this is the breadboard layout for the water sensor experiment in Chapter 11). The breadboard layouts in this book don't show the black lines that you can see in Figure 1-4, but they are shown in this photograph so you can see how the internal connections of the breadboard link the various component leads and wires together.

Once you become more experienced in electronics, you will probably want to learn how to make your electronic circuits more compact and permanent. You'll then need to learn how to build electronic circuits on *stripboard* or printed circuit boards by using solder and a soldering iron. But, for now, as you learn to create circuits, we'll use a breadboard method that allows you to correct mistakes easily.

Interconnecting Wires

 INTERESTING FACT

A *conductor* is a material that allows electricity to flow through it; an example is copper wire. An *insulator* is a material that does not allow electricity to flow through it; an example is plastic, such as the plastic insulating sheathing that surrounds the wire.

Along with the internal connections inside the breadboard, you will often need to create additional electrical connections to the components that you insert. To do this, you will need to use some solid *insulated copper wire* of a suitable diameter and

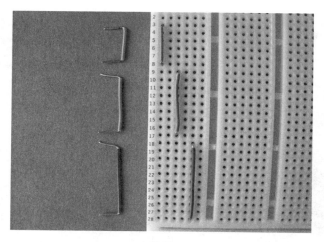

FIGURE 1-5 The interconnecting wires can be pushed into the breadboard holes.

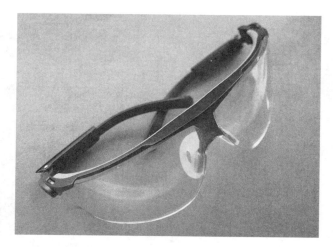

FIGURE 1-6 Safety glasses

current rating, which has been suitably *stripped* (that is, the ends of the wire have been stripped of the plastic insulation so that the copper wires are bare). The stripped part of the insulated wire can then be simply pushed into the holes like the component leads—an example of this is shown in Figure 1-5.

 NOTE

You should be able to purchase *precut* and *prestripped* interconnecting wires from the same electronic supplier that you purchase your breadboard from. However, you might find it more cost-effective to buy a roll of insulated solid copper wire from an electrical supplier and then cut and strip the wires to size yourself. You'll read how to do this shortly.

Safety Glasses

You should always wear safety glasses to protect your eyes if you decide to cut or strip interconnecting wires or component leads. Safety glasses come in many different shapes and sizes, and it is important that you choose a pair that fit your head and face correctly and cover your eyes sufficiently. My safety glasses are shown in Figure 1-6.

 BE CAREFUL!

Always wear safety glasses when cutting or stripping wires or leads. Also, make sure that you hold one end of a component lead when cutting it to keep it from flying through the air into your or someone else's eye. Also be aware that wire cutters and strippers are sharp and can cut into your skin.

Wire Cutters and Strippers

Wire cutters, like those shown in Figure 1-7, are useful for trimming interconnecting wires and can also be used to strip back the wire's plastic insulation.

Figure 1-8 shows you how you can use wire cutters to trim back, or strip, the plastic insulation off the wires. Simply hold the wire in one hand while carefully cutting into the insulation (but not into the metal wire) with the wire cutters. Then, carefully slide the insulation away with the cutters. Make sure that you practice the technique on some scrap pieces of wire before you start a project; it takes practice to get it right.

You can also use *wire strippers* to strip the insulation, and these make the job a lot easier.

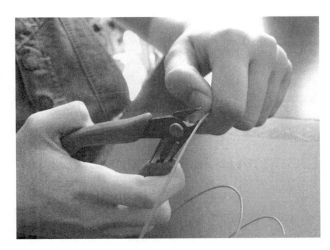

FIGURE 1-7 Wire cutters can be used to cut wires to size.

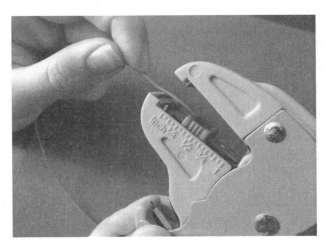

FIGURE 1-9 Position the wire into the jaws of the wire strippers.

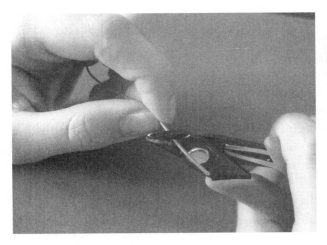

FIGURE 1-8 You can also use wire cutters to strip the insulation from interconnecting wires.

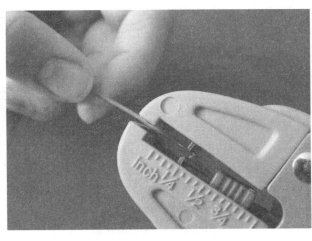

FIGURE 1-10 Squeeze the handle so that the jaws strip the insulation.

Wire strippers come in various shapes and sizes, and the type that I used to build the experiments in this book is shown in Figure 1-9. This photograph shows you how to strip the insulation from the wire: First, carefully insert the wire into the mouth of the strippers, and then squeeze the handle, as shown in Figures 1-9 and 1-10. This will produce a cleanly stripped cable. The strippers that I use include a wire cutter conveniently built into the handle, so I don't need to use a separate wire cutter to cut the wires to size; I can simply squeeze the handle of the wire strippers to cut the cable, as shown in Figure 1-11.

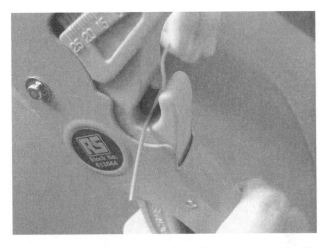

FIGURE 1-11 A handy cutting tool is sometimes built into the handle.

BE CAREFUL!

Always keep your fingers away from the stripping and cutting parts of the wire strippers. If you are not confident using wire cutters or strippers, ask a responsible adult to help you out.

Multimeter

A *multimeter* is a device used to take measurements in an electronic circuit, such as *voltage*, *current*, *resistance*, and *capacitance*. (You will learn about these measurements in Chapter 3.) A fairly basic multimeter that I have used for many years to build electronic circuits is shown in Figure 1-12.

This type of multimeter is sufficient for the projects in this book; it shows electrical measurements using a digital display. The price of multimeters varies from $10 to many hundreds of dollars, but you can get by with a basic multimeter to follow the experiments in this book; at a minimum, it should be able to measure voltage, current, resistance, and capacitance. You can also build the projects in this book *without* using a multimeter, but you'll get the most out of the experiments if you use one.

Take measurements by connecting the meter to the circuit using the two leads and probes—more

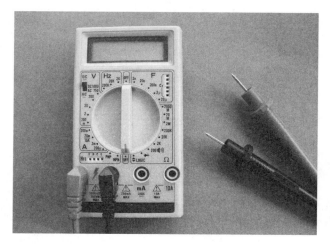

FIGURE 1-12 A multimeter should measure voltage, current, resistance, and capacitance.

details of how to do this are described in each chapter. Some of the more expensive multimeters are called "auto-ranging" meters, which means that the meter automatically adjusts the reading on the display. Otherwise, basic multimeters require you to adjust the meter setting manually.

Electronic Components

Without electronic components, we can't build an electronic circuit. Chapters 2 and 3 discuss some of the different types of electronic components used in this book in more detail. Some of the electronic components that you will be using are shown in Figure 1-13.

NOTE

The component descriptions and part numbers required for each project will be outlined clearly in the parts list in each chapter. The book's Appendix also suggests some useful resources to help you find components. Some experiments described in the book require you to use other household items as well, and these will be clearly outlined in each chapter.

Antistatic Precautions

Some sensitive electronic components, such as integrated circuits, can be damaged by static electricity. Static electricity can build up inside your body just by walking across a carpet, for example; have you ever gotten an electric shock when you touched another person or thing after walking across a carpet? This static electricity buildup can be transferred from your fingers into delicate electronic components, too, and it can sometimes damage them.

NOTE

You can learn more about electrical grounding and antistatic precautions on the Internet.

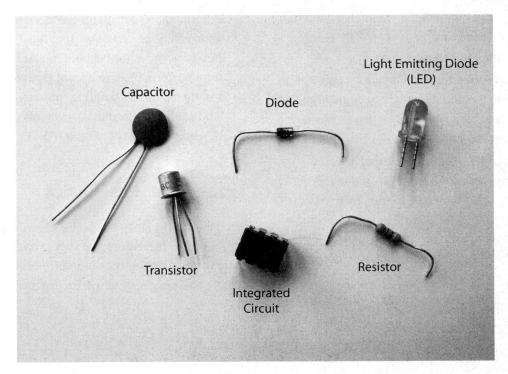

FIGURE 1-13 A selection of electronic components that you will be using

To remove static electricity from your body, you can wear an *antistatic wristband* with a strap and alligator clip connected to an *antistatic mat*; the mat is then connected to an *earth point* or *ground*, often supplied by plugging the mat into a three-hole electrical outlet. (In the United States, that third, round hole in the outlet is the *ground*. It's connected to some wiring in the wall that disappears into the ground. In the United Kingdom, earth grounding for antistatic mats is sometimes achieved via a special plug that links everything to the "earth pin" of your household electricity supply.) You should be able to purchase these items from an electronics supplier.

 **HINT!**

Always avoid touching the metal pins of an integrated circuit when building your experiments.

Warnings Before You Get Started

Playing and experimenting with electronics can be fun, but you should be aware of some important things before you get started:

● Always read through each chapter first before embarking on building the experiment, and then build the experiments only if you feel confident in your ability to do so.

● Some experiments include flashing lights. If you are affected by flashing lights or suffer from epilepsy, please do not build these experiments.

● Electronic components are small and can be a choking hazard. Make sure that you keep them away from small children.

● The components in these experiments should not become hot in normal operation. If any of the components become hot, remove the battery from the circuit immediately and check that the circuit has been constructed correctly before reconnecting the battery.

- Never short out a battery, i.e., never connect the positive (+) and negative (−) connections of a battery together. This can be dangerous and could cause the battery to leak or explode.

- Always connect electrolytic capacitors the correct way around in a circuit so that the positive lead is connected to the positive side of the circuit and the negative lead is connected to the negative side of the circuit. Connecting an electrolytic capacitor the wrong way can cause it to leak or explode.

- *Never* connect these experiments to the household electricity supply—*mains electricity* (the AC power in your house) can kill you.

- Each of the experiments has been tested extensively as part of writing this book; however, the author cannot guarantee the long-term performance, or accept legal responsibility for, the results of building these experiments. The reader builds the experiments outlined in this book at his or her own risk.

Ready for the Basics

You are almost ready to start building the experiments in this book. But before you start to build anything, you still need to understand some electronics basics. In the next chapter, you will do just that!

CHAPTER 2

Zooming into Electronics: Electronic Building Blocks

THIS CHAPTER WILL HELP YOU TO THINK like an inventor and an electronics design engineer. You will learn that electronic devices are made up of a number of different building blocks that interact with each other to create a finished product. You'll explore how electronic devices can be designed using various electronic building blocks, which sets the scene for the experiments in later chapters.

Think Like an Inventor

Every electronic device is created by design, using a number of individual *electronic components* that are linked together to produce an *electronic circuit*. The number of electronic components that are used to make up an electronic circuit will vary, depending on how complex the device is. Usually, the more complex the device, the more electronic components are required. You will be reading about electronic components in more detail in Chapter 3.

Lots of the fun of experimenting with electronics comes from imagining what you want to create, and then figuring out what you need to do to create it. Before you select any electronic components, though, you need to start thinking like an inventor and an electronics design engineer. A good inventor

or designer begins with a few preliminary steps to get her ready to build a creation. There are three main stages in the process:

1. Create a design specification.
2. List the building blocks.
3. Design and test the electrical circuit.

Write the Design Specification

Let's consider, for example, an inventor who wants to design and build a telephone. The inventor will start by writing down a list of things that she imagines the telephone will do. These features are also known as *specifications*. For a telephone, the main specifications might look like these:

- The device will enable two people at some distance away from each other to carry on a conversation.
- Each person will be able to talk into the telephone via some device.
- Each person will be able to hear the other person's voice via some device.
- Each telephone will be connected to another telephone via a cable.
- The audio signals from each telephone will be transmitted across a cable.

- Each telephone will be assigned a number that must be typed in by a caller to call that telephone.
- Each telephone must include some sort of device for indicating the number of the person to be called.
- Each person's phone will ring when another person is calling.

List the Electronic Building Blocks

Once the design specification has been listed, the inventor can then list the *building blocks* that are needed to create each of the specifications for the telephone. These might include the following:

- **Earpiece**   This will allow one person to hear the other person talking.
- **Microphone**   This will enable one person to talk to the other person.
- **Numeric keypad**   This allows the caller to punch in the other person's telephone number to make the receiver's phone "ring."
- **Display**   This allows the caller to see the number he has typed into the keypad.
- **Ringer**   This audible device will be positioned inside the telephone to make a noise to let the receiver know that someone is calling him and wants to talk.

Design and Test the Electronic Circuit

After the inventor has identified each of the building blocks required to make the telephone, she can use her knowledge of electronics and electronic components to begin designing individual electronic circuits that will perform the requirements of each building block. Each of these individual circuits will then be combined to produce one complete electronic circuit to create the functions of a working telephone. The inventor will then spend some time building and testing the final electronic circuit to make sure that it performs as she expects.

If the design process has been successful, the inventor will have created a working telephone that meets her original design specification. You can see from this simple example that a number of design stages are involved before an inventor actually picks up any electronic components and starts building.

Zooming into Electronics

Now imagine that you have the inventor's finished telephone in front of you, and you want to understand how it works. To do this, you'll use something called "reverse engineering," which means you will examine the creation to try to retrace the steps of the original inventor. You might imagine, for example, that you have a special microscope that allows you to examine the telephone on three different levels. The "special microscope" that allows you to do all this is your mind. As you become more experienced in understanding how electronic building blocks and circuits can be made, you will be able to construct (and deconstruct) electronic circuits more easily.

1X Magnification

At the first level, you'll look at the electronic device with your naked eye. You can see the electronic device, understand what it does, and see its basic features.

2X Magnification

At the second level, you are looking inside the telephone, where you can identify the individual building blocks that make up the telephone. For example, you can see that the telephone has a keypad that lets you type in your friend's telephone number, and it has a screen that allows you to see the telephone number that you have typed into the phone.

3X Magnification

Finally, this magnification level allows you to zoom even further into each building block to identify and examine the individual electronic circuits and components that make up each of the individual building blocks. You'll read more about components in Chapter 3.

Electronic Building Blocks

Now that you have learned that electronic circuits can be made up of building blocks, you might be wondering what these building blocks are. Take a look at the diagram in Figure 2-1, which shows four main electronic building blocks that can be used to make up any electronic circuit: power supply, input, output, and control circuitry.

These building blocks could be an individual electronic component, or they could be a number of electronic components linked together to form an electronic circuit. By learning about many of these building blocks used throughout this book, you will begin to discover how easily you can zoom in to spot them in everyday household electronic products.

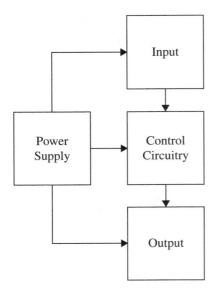

FIGURE 2-1 The four main electronic building blocks and how they interact

Power Supply

Every electronic circuit needs to have a *power supply*; without power, the circuit won't work! Electronic devices need electricity to work. Look at the block diagram in Figure 2-1, and you will see that the power supply is directly connected to each of the other three building blocks.

BE CAREFUL!

Do not attempt to use your household supply to power the experiments in this book. The household power supply can give you a nasty electronic shock and could kill you. That's why we'll be using low-voltage power supplies (batteries) for all the projects in this book.

Here are a few examples of power supplies:

- **Low-voltage battery**   The type of battery that you use in your TV remote control, flashlight, or smoke alarm is considered a low-voltage battery. The voltage of common household batteries varies and is typically around 1.5 to 9 volts. These batteries can be connected together to produce higher voltages.

- **High-voltage supply**   This type of power is supplied throughout your house by your household wall sockets. This high voltage is normally between 110 and 230 volts, depending on where you are in the world. This high voltage is usually reduced, using various components and circuitry, to a lower voltage level (such as 5 volts) to match the sensitive electronic components in the circuit.

- **Solar power**   This type of power uses sunshine that is converted into electricity.

- **Generator or dynamo**   This type of power is generated by a device that converts moving energy into electricity.

HINT!

You will be using low-voltage household batteries in all the experiments in this book.

Input

Inputs are electronic components (or individual circuits) that allow the outside world to interact with the electronic circuit in some way. Activating an input normally makes the electronic circuit do something.

Here are some examples of input building blocks:

- **Switches**   These could be individual toggles or switches (such as an on/off button), a keypad, or even a keyboard like the one you use with your computer.

- **Temperature sensor**   This device senses a change in temperature and could make the circuit do something if the temperature reaches a certain level.

- **Light sensor**   This device senses a change in light levels and makes the circuit do something if the light level reaches a certain level.

- **Infrared receiver**   This device senses infrared light (which is invisible to the human eye) to act as a switch in a circuit.

- **Microphone**   This converts a sound (for example, your voice) into an electronic signal.

Output

Outputs are electronic components (or circuits) that are the "doing" part of the circuit and make the circuit interact with the outside world in some way.

Here are some examples of output building blocks:

- **Light emitting diode (LED)**   This component converts electricity into light. You will soon see how you can use LED devices as indicators or illumination devices.

- **Speaker**   This converts an electronic signal into a sound that you can hear.

- **Buzzer**   This converts an electronic signal into a noise; this could be used as a warning device.

Control Circuitry

The *control circuitry* is the heart of the circuit that processes the signals from the input device and does something. Normally, this will produce a signal to an output building block. You will be learning about many different ways that you can create control circuitry in the following chapters.

How the Building Blocks Might Look

Figure 2-2 shows a circuit board layout of a project that I built a while ago (don't worry—the experiments in this book aren't as complicated as this).

The board layout looks complicated, but if you look at Figure 2-3, you can see that the layout can actually be broken down into a number of smaller electronic building blocks that are linked together to produce the final circuit.

FIGURE 2-2 An example of a typical electronic circuit

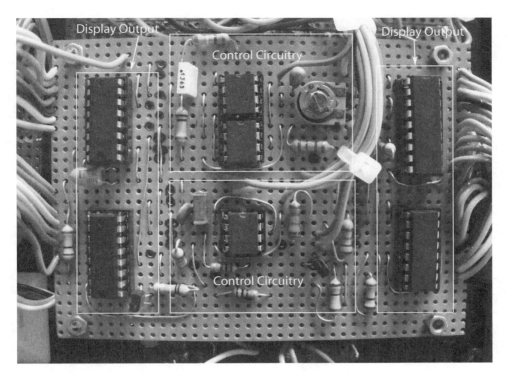

FIGURE 2-3 The complex circuit is built of a number of simpler building blocks.

Moving On

By reading this book, you will learn that you can link many simple electronic building blocks together to produce a more complex electronic system. Now read onto the next chapter, which explains the operation of some of the key electronic components that you will be using in this book.

CHAPTER 3

Examining Some Key Electronic Components

MANY DIFFERENT TYPES OF ELECTRONIC components are available to you as an electronics hobbyist, and each has its own particular characteristics. This chapter introduces you to some of the common electronic components.

To build electronic devices, you connect components together in a particular way to create an electronic circuit like the one shown in the *circuit diagram* in Figure 3-1. This type of diagram is also sometimes called a *schematic*.

A circuit diagram uses various circuit symbols that represent each of the electronic components used in the device; these components are linked together by lines that represent the electrical wires that join the individual components together. As you work through this book you will soon learn how to read and understand this type of circuit diagram. Schematic symbols are international symbols, and when you master them, you will be able to understand any device schematic created by anybody anywhere in the world. When you learn to read circuit diagrams, you're joining a special club.

Wires that are joined or attached together look like this on a circuit diagram:

Wires that are not joined together (overlapping wires) are shown like this on a circuit diagram:

Power Supply

Before you learn about some of the common electronic components, you need to understand the fundamentals of how electricity flows through an electronic circuit. If you read Chapter 2, you know that every electronic circuit needs a power supply to make it work; the power supply that you will be using in each of the experiments is a common low-voltage household battery, the same type that you use to power your TV remote control.

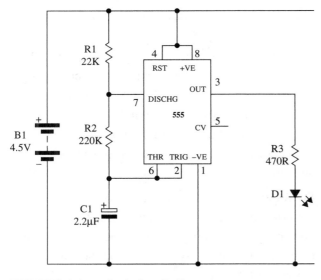

FIGURE 3-1 A typical circuit diagram

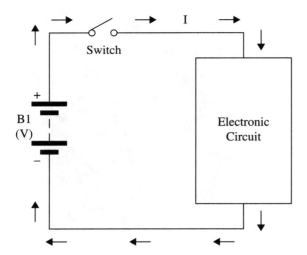

FIGURE 3-2 A basic circuit diagram showing the conventional flow of electricity

To understand how electricity flows in an electronic circuit, take a look at a basic circuit diagram shown in Figure 3-2. The diagram shows the circuit symbol for a battery on the left side (B1), which is connected to a rectangular block on the right side, which represents the electronic circuitry. The arrows show the direction of *current* flow through the circuit.

A battery power source is packed with *electrons*, tiny subatomic particles that are determined to escape. When a battery is connected to a circuit using some interconnecting wire, the electrons inside the battery are able to flow out of one side of the battery, around the electronic circuit, and then back to the other side of the battery. So long as this continuous electronic circuit is unbroken, the flow of electricity continues in a never-ending loop.

 INTERESTING FACT

Diagrams showing *conventional current flow* show electricity flowing from the positive side (+) to the negative side (–) of the battery. In reality, electrons flow from the negative side of a battery to its positive side. The apparent reason for why circuit diagrams show current flow from positive to negative is that in the early days of electricity being discovered it was assumed that current flowed this way, and circuit diagrams have remained this way ever since.

Two main measurements relate to a power supply: *current* and *voltage*.

The amount of electrons that flow through an electronic circuit determines the amount of *current* that flows. Current is measured in *amperes*, or *amps*, and the measurement is shown as a symbol *A*. The experiments in this book have a low-current consumption, drawing less than 100 milliamps (0.1 amp); 1000 milliamps equals 1 amp. You will see that electronic calculations and circuit diagrams identify current with the letter *I*, and the arrows on the circuit diagram identify the direction of the current flow.

Voltage is the value of a battery and is measured in *volts*. This is also known as the *potential difference*, because this equals the energy difference between two points. Volts are shown using the symbol *V* on circuit diagrams and in electronic formulae.

Famous Scientists

The measurement of current is the *ampere*, which was named after the French physicist Andre Marie Ampere (shown left) who was born in 1775. The measurement of potential difference is the *volt*,

which was named after the Italian physicist Alessandro Volta (shown right) who was born in 1745.

You will be using various voltages to power each of your experiments; these voltages are created by connecting multiple 1.5 volt (AA type) batteries together in a *series*, as shown in Figure 3-3. Each 1.5 volt battery has two different *poles*, one at each end, and these are marked as positive (+) and negative (–).

So, for example, if you connect two 1.5 volt batteries together in series, you can create a 3 volt power supply. And if you connected three 1.5 volt batteries together in series, you would create 4.5 volts of power.

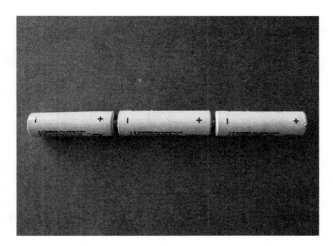

FIGURE 3-3 Batteries connected in **series** increase the overall voltage. Connecting in series means connecting the batteries end to end, with the positive (+) pole of one battery connected to the negative (–) pole of the next battery

The easiest way to connect batteries together like this to create different voltages is to use special battery holders that are prewired and allow you to fit batteries into them. A selection of battery holders is shown in Figure 3-4. You will be using battery holders in your experiments, which are easily obtainable from electronics supply companies, or you may be able to salvage some from broken electronic devices.

Switches

Switches let you either *make or break* an electronic circuit—that is, they let you either switch on or off the electricity flow in a circuit. Switches are available in various configurations and can be used to switch on or off your project or to make something happen. A selection of switches is shown in Figure 3-5.

Some switches are classed as *normally open*; this means that in the switch's normal state, the contacts inside it are open (not touching each other). When you press or activate the switch, the contact closes, which activates the circuit in some way. The circuit symbol for a normally open switch that could be used to switch a circuit on and off looks like this:

The experimental circuits in this book don't use this type of switch, but if you decide to make your experiments more permanent, you could include one of these switches just after the positive battery connection (shown in Figure 3-2), so that you can switch the circuit off when it is not being used.

The circuit symbol for a push-button switch, which could be used as part of a keypad, for example, looks like this:

FIGURE 3-4 Various battery holder configurations

FIGURE 3-5 A selection of switches

This type of switch is called a *momentary switch*, because the circuit closes when you press the switch and then opens when you take your finger off the switch.

Some switches are classed as *normally closed*; this means that in the switch's normal state, the contacts inside the switch are closed (touching each other), so when you press or activate the switch, the contact opens, which deactivates the circuit. You won't be using this type of switch in any of the experiments in this book.

Resistor

A *resistor*, like the examples shown in Figure 3-6, is a component that restricts the flow of electricity and can therefore reduce the amount of current that flows in a circuit. You might be wondering why you would want to restrict the amount of electricity flow in a circuit. One reason is because some components can be damaged if the current flow is too high—you will discover more about this as you read through the book.

A resistor has two leads, and its value is measured in *ohms*. The symbol for ohms is the Greek letter *Omega* (Ω). In electronic calculations, the letter *R* represents resistance.

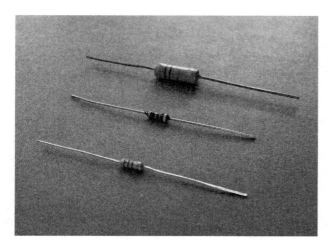

FIGURE 3-6 A selection of resistors

Resistors are available in many different values, from 1 ohm up to many millions of ohms. There is a range of resistors called *E12*, and these are available in 85 different values. The values of a resistor are normally shown in the following formats:

Kilo-ohms = KΩ   1KΩ = 1000 ohms
Mega-ohms = MΩ   1MΩ = 1 million ohms

Famous Scientists

The measurement of a resistor, *ohms*, is named after the German physicist Georg Simon Ohm who was born in 1787. Ohm also discovered the formula called *Ohm's Law*, which you will read about in Chapter 4.

Two types of circuit symbols can be used for a resistor in a circuit diagram, as shown here:

There are slight differences in some circuit symbols depending on which country you live in. In the preceding example of the resistor, both symbol styles represent the same type of component. The style on the left is used in the circuit diagrams in this book.

Calculating the Value of a Resistor

You can work out the value of a resistor using its color code; normally, four (sometimes five) color bands encircle the resistor, and you can calculate the value of the resistor by reading the color bands and working out their value.

Here's how to read the first three values of a resistor with four color bands:

- The first color band of a resistor denotes the "tens" value of the resistor.
- The second color band of a resistor denotes the "units" value.
- The third color band is the multiplier.

So, for example, if the first three bands of a resistor are brown, black, and red, you can see by looking at Table 3-1 that you take the values *1* (brown) and *0* (black) to make *10*. You then multiply this number by the third band multiplier, which is *100* (red). This gives us a value of 1000Ω, which is actually 1KΩ.

Finally, the fourth color band denotes the tolerance of the resistor:

Brown = ±1%   Red = ±2%
Gold = ±5%   Silver = ±10%

So, for example, if the fourth band of the example resistor is gold, the resistor has a manufacturer's tolerance of ±5%. This means that the value of the resistor could actually be

1KΩ + 5% tolerance = 1050Ω

Or it could be

1KΩ – 5% tolerance = 950Ω

If it is not easy to read the color code of a resistor, you can always check its value by using your multimeter. You can do this by setting your multimeter to a resistance setting, and then connecting the positive lead to one lead of the

FIGURE 3-7 Using the probes of a multimeter to check the value of a resistor

resistor and the negative lead to the other lead of the resistor, like the photograph in Figure 3-7.

It doesn't matter which way around you connect the leads, because the resistance can be measured either way around. If your meter is not an auto-ranging meter and you don't see a reading on the meter display, you need to switch the meter setting to a different resistance range until the meter's display shows the resistance value. You can see this in Figure 3-8, which shows a 4.7KΩ resistor being measured using a multimeter, and you can see that

TABLE 3-1 Resistor Color Codes for E12 Type Resistors with Four Color Bands

First Color		Second Color		Third Color (Multiplier)							
				x 0.1	x 1	x 10	x 100	x 1000	x 10000	x 100000	x 1000000
				Gold	Black	Brown	Red	Orange	Yellow	Green	Blue
Brown	1	Black	0	1Ω	10Ω	100Ω	1KΩ	10KΩ	100KΩ	1MΩ	10MΩ
Brown	1	Red	2	1.2Ω	12Ω	120Ω	1.2KΩ	12KΩ	120KΩ	1.2MΩ	
Brown	1	Green	5	1.5Ω	15Ω	150Ω	1.5KΩ	15KΩ	150KΩ	1.5MΩ	
Brown	1	Gray	8	1.8Ω	18Ω	180Ω	1.8KΩ	18KΩ	180KΩ	1.8MΩ	
Red	2	Red	2	2.2Ω	22Ω	220Ω	2.2KΩ	22KΩ	220KΩ	2.2MΩ	
Red	2	Violet	7	2.7Ω	27Ω	270Ω	2.7KΩ	27KΩ	270KΩ	2.7MΩ	
Orange	3	Orange	3	3.3Ω	33Ω	330Ω	3.3KΩ	33KΩ	330KΩ	3.3MΩ	
Orange	3	White	9	3.9Ω	39Ω	390Ω	3.9KΩ	39KΩ	390KΩ	3.9MΩ	
Yellow	4	Violet	7	4.7Ω	47Ω	470Ω	4.7KΩ	47KΩ	470KΩ	4.7MΩ	
Green	5	Blue	6	5.6Ω	56Ω	560Ω	5.6KΩ	56KΩ	560KΩ	5.6MΩ	
Blue	6	Gray	8	6.8Ω	68Ω	680Ω	6.8KΩ	68KΩ	680KΩ	6.8MΩ	
Gray	8	Red	2	8.2Ω	82Ω	820Ω	8.2KΩ	82KΩ	820KΩ	8.2MΩ	

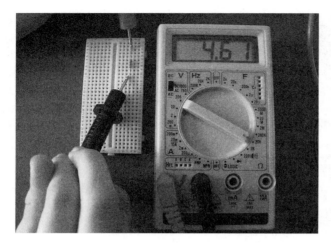

FIGURE 3-8 Make sure that the multimeter is set to read resistance.

FIGURE 3-9 A selection of variable resistors

in reality the meter is showing that the resistor has an actual value of 4.61KΩ. Note that the dial is set to point to 20K in the Ω section of the multimeter. This means that the meter is set for you to measure a resistor with a value up to 20KΩ.

Resistors are also available in various power (wattage) ratings, and it is important that the correct wattage rating is used. The required wattage rating of the resistors used in this book are outlined in the parts list in each experiment.

Variable Resistor

A *variable resistor* is also sometimes known as a *potentiometer* or a *preset*. It works like a resistor to restrict the flow of electricity, but the resistance value of the component can be altered from 0Ω to its maximum resistance value by you manually adjusting it. This can normally be performed by rotating the central part of the component using a flat bladed screwdriver. Other types of variable resistor have a metal or plastic shaft that allows you to rotate the central part of the component using your fingers. A selection of three different types of circuit board–mounted variable resistors is shown in Figure 3-9. The top line shows the tops of the resistors, and the bottom line shows the underside views of the same variable resistors; you

can see that normally this component has three connection leads: two on one end and the third on the opposite end.

Like fixed resistors, variable resistors are available in many different ohm and wattage ratings. You will have probably already used variable resistors at home—for example, these devices are used to change the volume on your home stereo.

Two types of circuit symbols can be used for a variable resistor in a circuit diagram, as shown here:

If you fit a 1KΩ variable resistor to a piece of breadboard and measure the resistance across the two end connections, as shown in Figure 3-10, your multimeter should read around 1KΩ (note that the actual meter reading is not shown in this photograph).

If you keep one end of the multimeter lead connected to one of the outer pins and connect the other multimeter lead to the center pin, you can see that adjusting the position of the central section of the variable resistor alters the value of the reading from 0 to 1KΩ. The photograph shown in Figure 3-11 shows that the central position of the variable resistor is at the 7 o'clock position and the meter shows a resistance reading of 0.7KΩ (which is 700Ω).

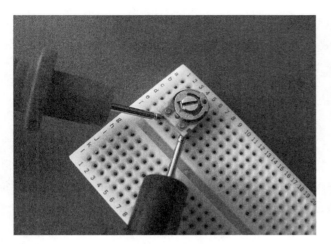

FIGURE 3-10 Measuring the two outer pins will show the overall resistance value on your multimeter (make sure that your meter is set to read resistance).

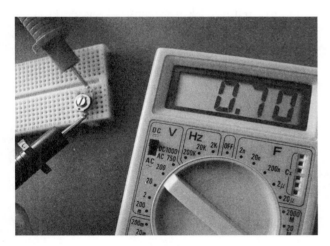

FIGURE 3-11 Adjusting the position of the spindle varies the resistance value.

Light-Dependent Resistor

A *light-dependent resistor (LDR)* works like a normal resistor, but its resistance value changes depending on the amount of light that shines on it. This type of device can be used in circuits to switch something when the light changes from light to dark, or from dark to light. They are often used in photographic light meters. Figure 3-12 shows a typical LDR.

FIGURE 3-12 A typical LDR

The circuit symbol for an LDR is shown here:

If you insert an LDR into a piece of breadboard, you can measure the resistance value of the device in the same way you would for a normal fixed resistor. You will also see how the resistance of the device changes whenever you shine a light onto the device or cover it up.

Capacitor

A *capacitor* is a component that can store an electrical charge and is used mainly to remove unwanted electronic signals in an electronic circuit; or it may be used in timing circuits.

A capacitor is basically two metal plates that are connected to two wire leads; the two plates are insulated from each other by a *dielectric*, which is normally a plastic film. This sandwich is then rolled up to produce a compact component.

Capacitors can be *polarized* or *non-polarized*. The polarized version is called an *electrolytic capacitor* and must be connected in the circuit so that the positive connection of the capacitor is connected to the positive battery leg of the circuit. Non-polarized capacitors can be connected in a circuit in any direction; it doesn't matter which way around it is connected.

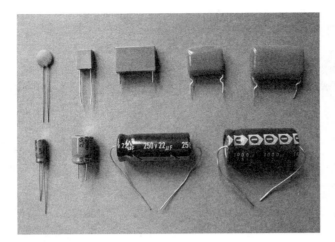

FIGURE 3-13 A selection of capacitors

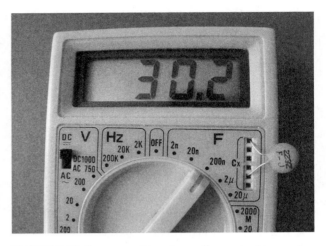

FIGURE 3-14 Some multimeters allow you to measure capacitance.

Figure 3-13 shows a selection of various capacitors and demonstrates the different shapes and sizes of these components. The top row are non-electrolytic (non-polarized) capacitors, and the bottom row are electrolytic capacitors.

The circuit symbol for a non-electrolytic capacitor is shown here:

Two types of circuit symbol can be used for an electrolytic capacitor in a circuit diagram, as shown next:

Capacitance is measured in farads (F), and the common values are as follows:

Picofarads = pF   1pF = 0.000000000001 farad
Nanofarads = nF   1nF = 0.000000001 farad
Microfarads = μF   1μF = 0.000001 farad

Famous Scientists

The measurement of a capacitor is *farad*, which is named after the physicist Michael Faraday who was born in 1791. Faraday experimented with electricity and proved the concepts for the operation of an electric motor.

The multimeter that I use has a connection that allows you to plug in a capacitor to measure its value. In Figure 3-14, for example, the meter is showing a capacitance of 30.2nF, though I was measuring a 22nF capacitor at the time. The reason for the difference between the rating and the multimeter measurement is that components have manufacturing *tolerances*, and this means that they are not always the exact value that is stated on the side of the component. In addition, variances in the temperature can cause changes in the readings, or the multimeter may not always be 100 percent accurate, and this can also contribute to the difference in the reading.

Some (though uncommon) non-electrolytic capacitors use a color code that is similar (but not the same) to the resistor color coding, and this allows you to identify the value of the capacitor. Electrolytic capacitors do have markings on them that help you identify which lead is positive and which is negative. Capacitors also have a voltage range, and you must make sure that the voltage rating of the capacitor value is *higher* than the voltage of the circuit.

Diodes

A *diode* allows electricity to flow in one direction but not the other, and this means that a diode can be used to protect certain parts of a circuit by blocking signals from flowing where they shouldn't. A selection of diodes is shown in Figure 3-15.

The circuit symbol for a diode looks like this:

One leg of a diode is called the *anode*, and this is normally connected to the positive side of a circuit; this is sometimes shown with a letter *A* next to it on a circuit diagram. The other leg is called the *cathode*, and this is connected to the negative side of a circuit and is sometimes shown with a letter *C* or a *K* next to it on a circuit diagram. If a diode is connected in a circuit this way, it is said to be in a *forward bias*, which means that current *will* flow. If the polarity of the battery is reversed, it is said to be *reverse biased*, which means that current *will not* flow.

A diode also has a small voltage drop, which varies from 0.4 to 0.7 volts, depending on whether the diode is made from silicon or germanium.

INTERESTING FACT

There is another type of diode called a *Zener diode,* as shown in middle of the photograph in Figure 3-15. These devices are manufactured to have a specific voltage drop rating and can be used in circuits where you want something to happen whenever a specific voltage is reached; or they can also be used to create a specific voltage in a circuit. Zener diodes are not covered in this book.

Your multimeter should have a diode setting, which is normally identified by the circuit symbol for a diode. Figure 3-16 shows how you can use your multimeter to test the volt drop of a standard diode.

If you were to connect the multimeter probes the other way around, the multimeter would not show any reading, because the diode conducts electricity in one direction only.

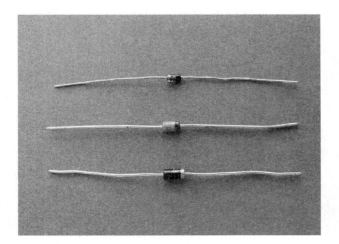

FIGURE 3-15 A selection of diodes. The band around a diode identifies the cathode lead.

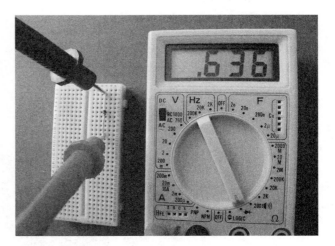

FIGURE 3-16 Testing a diode with a multimeter

Light Emitting Diode (LED)

A *light emitting diode (LED)* is similar to a standard diode in that it allows current to flow in one direction only; the difference between the two is that the LED also produces a light output when it is forward biased. Various sizes, shapes, and sizes are available, and some display packages contain many LEDs to create number and letter shapes. You may have seen a seven-segment LED display in a digital alarm clock. A selection of LEDs is shown in Figure 3-17.

The circuit symbol for an LED looks like this:

The symbol looks similar to the symbol for a standard diode, with the addition of the arrows, which indicate that the diode illuminates.

 BE CAREFUL!

To avoid damaging an LED, you should always wire a resistor in series with it, and make sure that it is connected the correct way around in the circuit. This is explained in more detail in the experiment in Chapter 4.

Figure 3-18 shows a standard 5mm LED that you will be using in the experiments in this book.

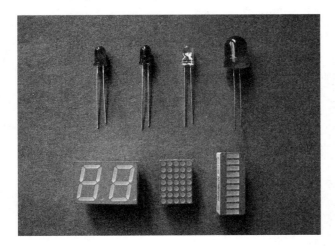

FIGURE 3-17 A selection of LEDs

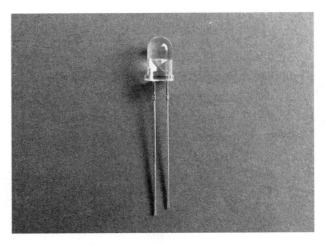

FIGURE 3-18 A typical 5mm LED

The longer lead of the LED is normally the anode (+) connection, and this is connected to the smaller electrode inside the LED. The shorter lead is the cathode (–) connection, and this is usually connected to the larger electrode inside the LED. The shorter cathode (–) lead is also usually positioned next to the flat edge of the LED.

Transistor

A *transistor* is basically an electronic switch that allows you to switch a large current by applying a smaller current to the device. They are often used inside an audio amplifier in a home sound system, where they can be used to convert a small sound signal into a much louder one. Transistors come in two main types, *NPN* and *PNP*, and both types have three pin connections, *base*, *collector*, and *emitter*. Figure 3-19 shows the BC108 and BC178 transistors that are used in the experiments in this book and also the position of its component leads compared to the tab on its metal casing.

The circuit symbol for an NPN transistor looks like this:

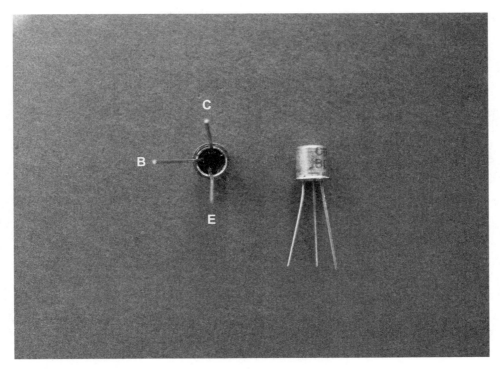

FIGURE 3-19 The type of transistor used in this book has a metal tab on its casing to help you to identify its leads; the **B** marks the base, **C** marks the collector, and **E** marks the emitter which is next to the metal tab.

And the circuit symbol for a PNP transistor looks like this:

If you apply a small signal to the base connection of the transistor, you can *amplify* this signal through the collector and emitter pins. This is explained in more detail in the relevant experimental chapters.

The measure of amplification of a transistor is called *hFE*—so, for example, a transistor with an hFE of 300 means that it is able to amplify the base signal by up to 300 times. Transistors come in various shapes and sizes, and each has its own application; some are suited to high-current applications and may be able to connect

to a *heatsink* (which transfers heat to reduce the component's temperature) to cool the device down in the circuit, and others can be used for general switching purposes. Some multimeters have a connection that allows you to test transistors, but basic multimeters may not have this feature.

Integrated Circuits

Integrated circuits (ICs) are small, rectangular packages that hold miniature electronic circuits inside, which could include many tiny capacitors, resistors, and transistors. The miniature internal circuitry allows you to reduce the amount of external components that you need to create a working circuit and helps to reduce the size of the overall circuit. A vast range of ICs are available and can be used in many different applications, such as for timers, LED drivers, sound effects, and audio

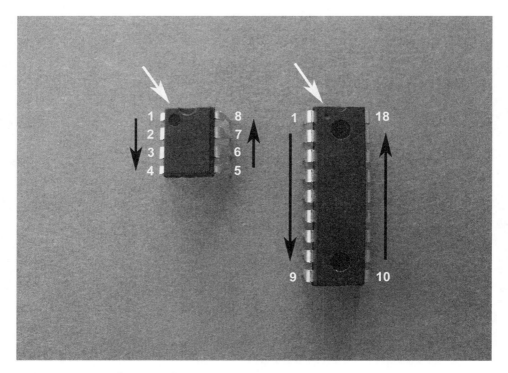

FIGURE 3-20 A selection of various size ICs showing the direction of the pins

amplifiers, to name a few. If we didn't have ICs, the electronic products that you use every day would be much larger in general.

ICs normally come in packages that have 8, 14, 16, 18, or sometimes more pins that allow you to connect it to a circuit. The pin numbers of an IC can be easily identified by looking for a circle or a semicircle that identifies the top of the IC. Pin number 1 is the upper-leftmost pin (shown with white arrows in Figure 3-20) and the pin numbers then follow the direction of the black arrows shown in the figure. ICs with 8, 14, 16, and 18 pins follow this format.

Sometimes ICs don't use a circle to identify pin 1, but once you have identified the top of the IC, the pin numbers are easy to figure out. ICs don't have a specific circuit symbol; they are normally shown on a circuit diagram as a rectangular block with connections coming out of it, like the *555* IC shown in the middle of Figure 3-1 at the beginning of this chapter.

To give you an idea of what the inside of an integrated circuit looks like, have a look at Figure 3-21, which shows an integrated circuit that contains a clear circular window through which you can see the miniature circuits inside.

See how the fine wires that are linked to the tiny circuit disappear off inside the IC? These are connected to each of the ICs pins, which you connect to various parts of the circuit. You can

FIGURE 3-21 Take a look inside an IC.

see that you would need a microscope to see how the tiny circuit is constructed, and if you did this, you would find that it contains many tiny components that are linked together. It's amazing how IC manufacturers can make circuits this small; imagine how big this circuit would be if we constructed it using normal sized components!

INTERESTING FACT

Here are some examples of how consumer electronics have shrunk in size over the years:

- **1930s** Cabinet-sized radios became available and used valves, or vacuum tubes, (a predecessor of the transistor) to amplify sound.
- **1950s** The first computers were built; they were huge—the size of an entire room—and used many valves.
- **1960s** Pocket-sized radios were available and used transistors.
- **1970s** Pocket-sized calculators were produced using integrated circuits.
- **1980s** Personal computers became available; some were as small as a shoe box.
- **1990s** Second generation (2G) mobile phones became commercially available. Earlier mobile phones were about the size of a brick; 2G phones began the move toward smaller devices.
- **2000s** Matchbox-sized personal audio players became available and used ICs to store many thousands of song tracks using a compressed digital data format (such as MP3).

Converting Circuit Diagrams to Breadboard Layouts

Each experiment in this book contains a circuit diagram that uses the component symbols that are shown in this chapter. The operation of the circuit is explained in detail to help you understand how it works.

Each experiment will show you how to build the circuit on a piece of breadboard; you don't need to convert the circuit diagrams into breadboard layouts, however, because this has already been done for you. By the time you have built the last experiment in the book, you should find it fairly easy to convert your own circuit diagrams into breadboard layouts. If you use a breadboard that's slightly different from the one that I use in the experiments, you should find it fairly easy to convert the book's breadboard layouts to suit your breadboard. Figure 1-3 in Chapter 1 shows how the internal connections of a typical breadboard are configured.

Experiment Building and Testing Guidelines

When you build the experiments in each chapter, follow the circuit diagrams and the close-up photographs of the breadboard layouts. Watch for some common errors when you're building the breadboard layouts, which are outlined here. Work through this checklist before you connect the battery to the circuit:

- Make sure that the battery leads are connected the correct way around in the circuit, as shown in the chapter photographs. Normally the red battery lead identifies the positive (+) battery connection and the black battery lead identifies the negative (−) battery connection.
- Are the batteries fresh? Old, depleted batteries may not have enough electricity to operate your experiment.
- Are you using the suggested battery voltage outlined in the experiment?
- Make sure that the component leads are connected the correct way around—for example, transistors, LEDs, electrolytic capacitors, and ICs must all be connected the correct way; otherwise, they may not work and could be irreparably damaged.

PUBLIC LIBRARY
DANVILLE, ILLINOIS

- Are you using the component values that are specified in the parts list? If you are using alternative component values, this could make the circuit operate in an unusual or erratic way.

- Check that you have not missed any of the wire links that connect various components together.

- Make sure that the components are fitted snugly into the breadboard; otherwise, they may not be making the correct electrical connection.

If you find that the circuit does not work as expected, don't panic; just immediately remove the battery from the circuit and check your layout again. You may have damaged a component or you may be lucky; you will need to check your layout again and if necessary alter some of the components if they are damaged. Fault-finding can be frustrating, but it's worthwhile in the end after you have a working circuit.

The best advice that I can give you is to take your time when building the experiments, making careful note of the images and recommendations in each chapter. If you do this, you should be able to get your experiments working the first time.

Time to Play!

A few other electronic components have not been described yet, but you will come across these in the experiments ahead and they will be explained in more detail in each project chapter. Now that you have learned about the key electronic components used in this book, it is time for you to play! So read on to the next section of the book, which shows you how to experiment with some telephone circuits.

PART TWO

Can You Hear Me Now? Phone Experiments

CHAPTER 4

Getting Charged Up! Make an LED Illuminate

In this part of the book, you'll learn how to build some experiments that are related to the mobile phone. Mobile phones are part of most people's everyday life; they come in many different shapes and sizes, and it seems that everybody owns one. Before you start to build your first experiment, take a good look at a mobile phone. What do you see? What features does it have? In other words, what can this mobile phone do? Think carefully, because some of its features will not be immediately obvious.

At the very least, your mobile phone will have a keypad so that you can type in telephone numbers or text messages. It will almost certainly have a screen that allows you to see the number that you are phoning or the name of the person calling you. The screen might also be quite large, and it might actually double as a keypad.

Features of a Typical Mobile Telephone

Table 4-1 describes some of the key features of a typical mobile phone. (The features that we are going to explore are identified by an asterisk.) Consider how these features can also

be categorized into one of the four main building blocks that you read about in Chapter 2.

You might see more features in the table than you expected, or you may be able to think of some other features that are not listed. Each of these individual features is designed by the phone manufacturer to be connected together to create the complete mobile phone that you carry around with you in your pocket every day.

The experiments that you are about to encounter in this part of the book show you how to use various electronic components to emulate some of the features of a mobile phone.

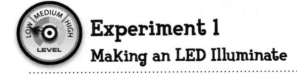

Experiment 1
Making an LED Illuminate

This experiment shows you how to illuminate a light emitting diode (LED) that emulates the charge indicator on a mobile phone. This indicator on a phone is designed to show you that the mobile phone's battery is being charged. An LED indicator is probably one of the easiest electronic circuits that you can build, and it also demonstrates how voltage, current, and resistance relate to each other in an electronic circuit.

TABLE 4-1 Features of a Mobile Phone

Feature	Description	Type of Building Block
Keypad	Allows you to type phone numbers and messages	Input
Color display*	Lets you see text and images	Output
Touch screen	Lets you control the phone by pressing icons on the screen	Input
Microphone	Converts your voice into an electrical signal	Input
Speaker	Converts an electrical audio signal into noise	Output
Bluetooth	Provides wireless connectivity links to other phones and computers	Input and output
Aerial	Allows communication to a mobile network	Input and output
Camera lens	Lets you take photographs	Input
LED charge indicator*	Shows that the phone is being charged	Output
LED message indicator*	Shows that a text message has been received	Output
Headphone socket	Accepts a headphone plug so you can listen to music	Output
Music player	Allows you to store and play back music files	Control
Video player	Allows you to store and play back video files	Control
Voice recorder	Converts noise into a audio signal and stores it as a file	Control
Contact list	Stores alphanumeric characters into memory	Control
Internal memory	Stores text, music, and video files	Control
External memory	Stores text, music, and video files (for example, a micro SD card)	Control
Telephone*	Transmits and receives radio signals and converts them into electrical audio signals	Control
Battery	Provides power to the mobile telephone	Power supply

BE CAREFUL!

Do not attempt to connect this or any other circuit discussed in this book to the household mains supply. The power supplied by mains can seriously hurt you. You will be using only low voltage batteries to power your experiments.

An LED is a versatile electronic component that converts electrical voltage into light output. LEDs come in various shapes, sizes, and colors and are used in many different display and indicator applications. They are popular because they use very little power and produce very little heat when they are illuminated.

NOTE

LEDs and other components were discussed in Chapter 3.

The Circuit Diagram

Take a look at the circuit diagram for this experiment, which is shown in Figure 4-1.

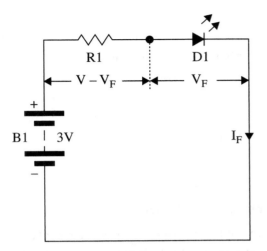

FIGURE 4-1 The circuit diagram for the LED indicator

If you are familiar with circuit diagrams, Figure 4-1 should be fairly easy for you to follow. If you're not, don't worry, because its operation is described in more detail shortly.

How the Circuit Works

This circuit is made up of two building blocks: power supply and output. The circuit is powered by two 1.5 volt AA batteries that are connected together in series to produce a 3 volt power supply. This is identified as B1 on the circuit diagram. The LED (D1) and the resistor (R1) create the output part of this circuit.

Here's how the circuit works: When the battery (B1) is connected to the circuit, electricity flows through the resistor (R1), which is included to limit the current that flows through the circuit. The electricity then illuminates the LED (D1) and flows back to the battery (B1). The LED remains illuminated as long as the battery is connected to the circuit.

 HINT!

It is important that you include a current-limiting series resistor when powering a standard LED; otherwise, you will damage the component.

LED Resistor Calculations

The formula for calculating the value of an LED series resistor uses Ohm's Law as its basis and this is shown next, you will notice that each of these parameters is identified on the circuit diagram in Figure 4-1.

 INTERESTING FACT

Georg Simon Ohm (1787–1854) discovered a relationship between resistance, voltage, and current. He created a formula to demonstrate this relationship, called *Ohm's Law*.

Ohm's Law states that

Resistance (R) = Voltage (V) / Current (I)

The formula for calculating an LED series resistor is very similar to that of Ohm's Law:

$$R = (V - V_F) / I_F$$

- **R** is the value of the series resistor (measured in ohms, Ω)
- **V** is the voltage of the circuit (measured in volts)
- V_F is the typical forward voltage drop of the LED (measured in volts)
- I_F is the amount of current that flows through the LED (measured in amps)

The values of V_F and I_F can normally be found on the LED manufacturer's datasheet. The amount of current that flows through an LED must not exceed the value of I_F as stated on the LED datasheet; otherwise, you could damage the LED. Typically, a 5mm red LED will have a forward voltage drop (V_F) value of around 1.8–2.8 volts and a maximum current rating of 20mA. The LED that I used in the experiment has a V_F rating of 2 volts, and I decided that I wanted to limit the current flow through the LED (this is the I_F value) to 10 milliamps (mA).

HINT!

Most LEDs will illuminate well with as little as 2–10mA current flowing through it. Reducing the current flow in an electronic circuit will make your batteries last longer.

So, for example, in this circuit, we can calculate the value of resistor R1 by substituting the values of our components into the formula:

$$R = 3V - 2V / 0.01A$$

Therefore, R = 100Ω

HINT!

You need to divide 10mA by 1000 to convert it to amps. This is why the I_F value is 0.01A in the calculation.

This shows us that we could use a resistor with a resistance as low as 100Ω to limit the current flowing in the circuit to 10mA. We are going to be using a resistor value of 470Ω in this experiment, which, by switching the formula around, shows that this will limit the current to the following:

$$I_F = (V - V_F) / R$$
$$I_F = (3V - 2V) / 470Ω$$

Therefore, I_F = 0.0021A

Multiplying the answer by 1000 shows that the current flow is 2.1mA. This is only an approximate value, however, because there are manufacturing tolerances of both the LED and the resistor. You will soon see for yourself what the actual current flow is in your circuit.

Things You'll Need

The components and equipment that you will need for this experiment are outlined in the following table. It's a good idea to find and prepare the items that you need before starting any experiment.

Code	Quantity	Description	Appendix Code
D1	1	5mm red LED	4
R1	1	470Ω 0.5W ±5% tolerance carbon film resistor	—
B1	1	3V battery holder	14
B1	2	1.5V AA batteries	—
B1	1	PP3 battery clip	17
—	1	Breadboard	1
—	—	Wire links	—
—	—	Digital multimeter	—
R1*	—	Various 0.5W ±5% tolerance carbon film resistors with values greater than 180Ω	—

*See the "Further Experimentation" instructions at the end of this chapter.

NOTE

The Appendix Code column of the table refers to specific parts that I used in this experiment. Details for sourcing these parts are outlined in the Appendix.

The Breadboard Layout

Follow the circuitry shown in Figure 4-2 carefully and build the circuit on a piece of breadboard. Plug the components and wire links into the breadboard.

NOTE

Refer to Chapter 3 for building breadboard layouts and fault-finding guidelines.

Notice how the flat side of the LED is positioned: this normally identifies the cathode (−) lead of the LED. You can see that this is connected to the negative lead of the PP3 battery clip with the black wire.

Time to Experiment!

After you've built the circuit and checked it over, insert the two AA batteries into the battery holder and connect it to the battery clip. If everything is connected correctly, you should find that the LED illuminates as shown in Figure 4-3.

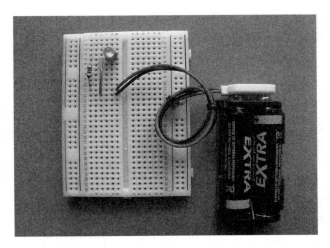

FIGURE 4-3 The LED is illuminated!

If the LED doesn't illuminate, remove the battery from the circuit immediately and check that your layout matches the one shown in Figure 4-2. It could be that you have connected the LED or the battery the wrong way around, in which case you could have damaged the LED—or you may have been lucky! Check your breadboard layout over again and make any necessary modifications

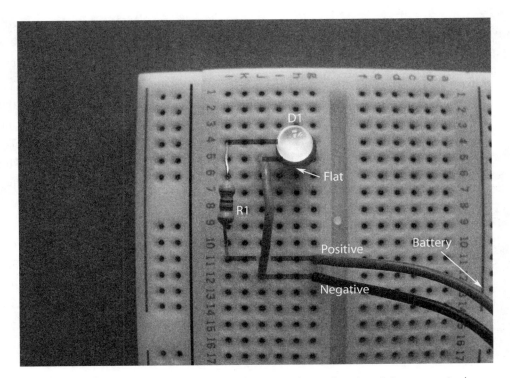

FIGURE 4-2 The gray horizontal lines show you how the circuit is connected together inside the breadboard.

before you try it out again. If the LED still doesn't illuminate, you may need to replace it with a new one.

Follow the circuit diagram in Figure 4-1 while looking at the breadboard layout that you have built. You should start to see how the circuit relates to the final layout.

Measuring Voltage

If your circuit is working OK, switch your multimeter on and select the DC voltage setting to enable you to measure at least 3 volts. Connect the positive and negative probes of the multimeter across the battery (in parallel), as shown in Figure 4-4, and your multimeter should read about 3 volts. You can see that my batteries are fresh, because the meter actually reads 3.21 volts.

Now connect your probes in parallel with the LED to measure the voltage across it, as shown in Figure 4-5. This is the voltage drop (V_F) across the LED, and its value depends on the type of LED that you are using. It should be roughly the value that we used in our formula earlier—around 2 volts. You will see that my meter reading actually shows 1.91 volts.

Now measure the voltage across the resistor, as shown in Figure 4-6, and you should see that it equals the difference between the battery voltage and the LED volt drop ($V - V_F$). In my circuit,

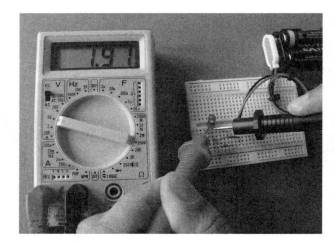

FIGURE 4-5 Measure the voltage across the LED.

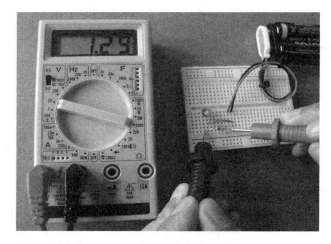

FIGURE 4-6 Measure the voltage across the resistor.

this was 1.29 volts, which can be calculated by subtracting the LED volt drop (1.91 volts) from the battery voltage (3.2 volts), which equals 1.29 volts. The readings in your circuit might be slightly different, depending on the battery voltage and the type of LED that you use.

Measuring Current

Now you are going to measure the current that is flowing through the circuit, which in this circuit is the value of I_F. To do this, you need to wire the multimeter in series with the circuit; *in series* means breaking into the circuit. But before you do this, remove the multimeter probes from the circuit and change the setting on your meter to read DC current.

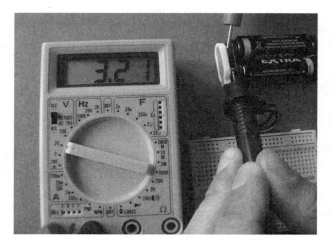

FIGURE 4-4 Measure the voltage of the battery.

You will probably also need to swap one of the probes over to another connection on your multimeter. (Refer to the instructions that came with your multimeter to find out how to do this.) This time, remove the positive battery connection from the breadboard and move it to an empty set of pins, and then insert the multimeter into the circuit between the positive battery connection and R1, as shown in Figure 4-7. Notice that my multimeter switch is set to read current up to 200mA (milliamps). My meter shows that only 2.7mA are flowing through the circuit; the current flow in your circuit may differ depending on the type of LED that you use.

If you substitute the readings that I made into the LED resistor formula, you can see how this works:

$$I_F = (V - V_F) / R$$
$$I_F = (3.21V - 1.91V) / 470\Omega$$
$$I_F = 0.0027A$$

Therefore, $I_F = 0.0027A \times 1000 = 2.7mA$

Notice that this is close, but not exactly the same, as the calculation that we made earlier. We originally calculated that the current flow was going to be 2.1mA. Don't forget that *tolerances* in the LED, the battery, the resistor, and your multimeter mean that there will be slight differences when you compare your calculated values against the readings that you actually make.

HINT!

The resistor that you used has a tolerance of ±5 percent; this means that the actual resistance could be as low as 446.5Ω or as high as 493.5Ω. You can measure the resistance of your resistor by following the instructions in Chapter 3 to see how this affects the results of your calculations.

Further Experimentation

Try changing the resistor (R1) for another with a higher resistance value, such as 2.7KΩ, and repeat the experiment to see what effect this has on the voltage and current readings.

BE CAREFUL!

Make sure that you alter the multimeter settings and the probe connections to the correct settings before connecting the probes to the circuit. Do not try to measure voltage with the multimeter set to read current, because this will either blow a fuse in your multimeter or damage it.

Now try a resistor with a lower resistance value, such as 180Ω, and see what effect this has. Do not use a resistor with a value that is lower than 180Ω in this experiment. You should also start to notice a relationship between the resistance value and the brightness of the LED. Try substituting the resistor value into the LED series resistor formula outlined earlier in the chapter to calculate the current flow, and then see how close this is to the readings that you see on your multimeter.

By experimenting, try to discover the answer to the following questions:

- What happens to the current flow in the circuit if you increase the value of R1?

- Does the LED shine brighter when the value of R1 is 470Ω or 1KΩ?

FIGURE 4-7 Measuring the current flowing through the circuit

- What effect does altering the value of R1 have on the $V - V_F$ value?
- What effect does altering the value of R1 have on the V_F measurement?

Summary

In this experiment, you learned how to illuminate an LED and how Ohm's Law works in this type of circuit. You also learned how the value of the series resistor affects the brightness of the LED.

This type of circuit could be used as a basic LED indicator module, and you might decide to use a different color LED instead of a red one, or an ultra-bright white LED that can be used to convert this circuit into a flashlight. Now read on to the next chapter, which shows you how to make an LED flash on and off.

CHAPTER 5

You Have a Message! Make an LED Flash

In Chapter 4, you examined some of the features of a mobile phone and learned how to make an LED illuminate just like the charge indicator on a phone. Another useful phone feature is an LED indicator that flashes when you receive a text message or a phone call. The experiment in this chapter shows you one method of making an LED flash on and off using an inexpensive integrated circuit that requires only four other components.

How to Make an LED Flash

To make an LED flash on and off, you need to build a circuit that makes an electrical signal switch on and off. This type of circuit is known as a *clock* or *astable circuit*, and you can build one in many different ways. One of the easiest methods is to use a popular integrated circuit (IC) called the "555 timer," and this is what you will be using in Experiment 2 in this chapter. A 555 timer IC has only eight pin connections and can be configured in two different modes: *astable* and *monostable*. This experiment explains how to build an astable circuit; the 555 timer's monostable operation is described in future chapters.

 INTERESTING FACT

The 555 timer has been available since the early 1970s and due to its low cost and versatility it still remains a very popular integrated circuit today.

 NOTE

You can read more about ICs and other components in Chapter 3.

First you need to learn about what an astable circuit is and what the electrical output signal looks like. Take a look at the diagram in Figure 5-1, which shows an astable waveform.

NOTE

A waveform shows what the voltage signal looks like in an electronic circuit. You use a special piece of measuring equipment called an *oscilloscope* to see a waveform.

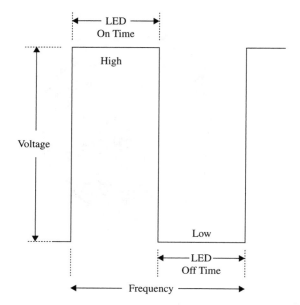

FIGURE 5-1 The output of an astable circuit looks like this.

This type of waveform is called a *square wave* and it shows how an electrical signal from an astable circuit switches on and off. The "high" part of the waveform represents a high voltage; normally this is the same as or almost the same as the battery

voltage of the circuit. The "low" signal represents a low voltage, and normally this is 0 volts. This signal repeats itself in a never-ending sequence and produces an output signal that turns on and off and on and off and on and off…. Imagine that this output signal is connected to an LED, and you can see that it will cause the LED to switch on and off at a regular rate.

Look at the basic circuit diagram for a 555 timer that is configured in astable mode, as shown in Figure 5-2.

The 555 timer IC has eight pin connections, and you can see from the circuit diagram that nearly all of the pins are connected into the circuit. Pin 3 is the output connection and produces the "on-off-on-off" electrical signal that looks similar to the square wave shown in the corner of Figure 5-2. You will be connecting this output to an LED in the experiment that follows.

The speed of the output signal can be controlled and is configured by altering the values of the three components that you see in Figure 5-2: the two resistors (R1 and R2) and the capacitor (C1).

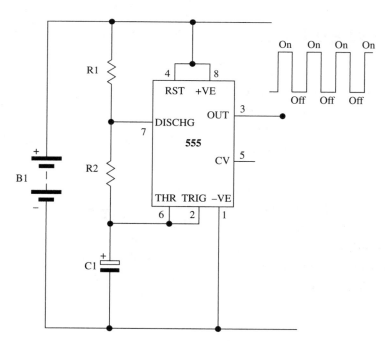

FIGURE 5-2 A basic 555 astable timer circuit and the output waveform

Before you build the experimental circuit, it is worth understanding how you can calculate the "On" and "Off" timings of this type of astable circuit. A few key formulas are associated with this mode of operation, and these are outlined here. Read these formulas in conjunction with Figures 5-1 and 5-2:

- The **on time** of the timer output (high output) is calculated by using this formula:
  On (measured in seconds) = 0.693 × (R1 + R2) × C1

- The **off time** of the timer (low output) is calculated by using this formula:
  Off (measured in seconds) = 0.693 × R2 × C1

- The **total time** period for one cycle (T) is therefore
  T1 + T2

- The **frequency** of oscillation (F) is expressed in Hertz (Hz):
  F = 1 / T

We will be using these formulas shortly.

NOTE

Frequency is the number of times T1 + T2 occurs every single second.

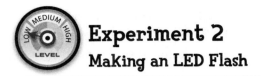

Experiment 2
Making an LED Flash

This experiment shows you how easy it is to use an astable 555 timer IC as a *control* building block to enable you to flash an LED on and off. You will also be experimenting with various component values to see what effect this has on the way that the LED flashes on and off.

BE CAREFUL!

This experiment shows you how to make an LED flash. If you suffer from epilepsy or are affected by flashing lights, this experiment is not for you!

The Circuit Diagram

Now that you understand how to make an astable circuit, take a look at the circuit diagram for this experiment, which is shown in Figure 5-3.

The circuit diagram looks very similar to the diagram in Figure 5-2, except it now includes some specific component values for R1, R2, and C1,

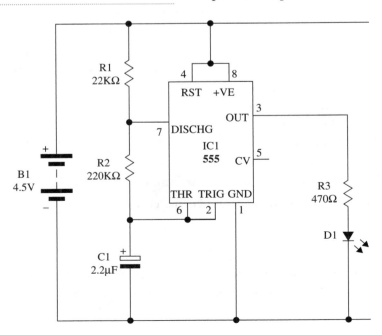

FIGURE 5-3 The circuit diagram for the LED flasher

which creates a suitable clock output on pin 3 of the 555 timer (IC1). Pin 3 is also connected to an LED (D1) via a series resistor (R3).

NOTE

Read more about calculating LED series resistors in Chapter 4.

How the Circuit Works

This circuit is made up of three building blocks: power supply, control circuitry, and output. This circuit is powered by three 1.5 volt AA batteries that are all wired in series to generate a 4.5 volt supply (B1). The 555 timer (IC1) and its associated timing components R1, R2, and C1 create the control circuitry in this experiment. The LED (D1) and resistor (R3) form the output part of this circuit, and if you have built the experiment in Chapter 4 you will have learnt about this building block already.

Having understood the operation of the 555 timer (IC1) as described earlier, you should be able to figure out how the circuit works already. The only addition to the 555 astable timer circuit is the LED (D1) and its series resistor (R3), which allow you to see the astable waveform output from pin 3.

The Breadboard Layout

Using the breadboard, build the circuit following the layout shown in Figure 5-4.

Things You'll Need

The components and equipment that you will need for this experiment are outlined in the following table. Locate and prepare the items that you need before starting the experiment.

Code	Quantity	Description	Appendix Code
IC1	1	555 timer	18
R1	1	22KΩ 0.5W ±5% tolerance carbon film resistor	—
R2	1	220KΩ 0.5W ±5% tolerance carbon film resistor	—
C1	1	2.2µF electrolytic capacitor (minimum 10 volt rated)	—
D1	1	5mm red LED	4
R3	1	470Ω 0.5W ±5% tolerance carbon film resistor	—
B1	1	4.5V battery holder	15
B1	3	1.5V AA batteries	—
B1	1	PP3 battery clip	17
—	1	Breadboard	2
—	—	Wire links	—
R1, R2*	—	Various 0.5W ±5% tolerance carbon film resistors	—

*See the "Further Experimentation" instructions at the end of this chapter.

NOTE

The Appendix Code column of the table refers to specific parts that I used in this experiment. Information about sourcing these parts is outlined in the Appendix.

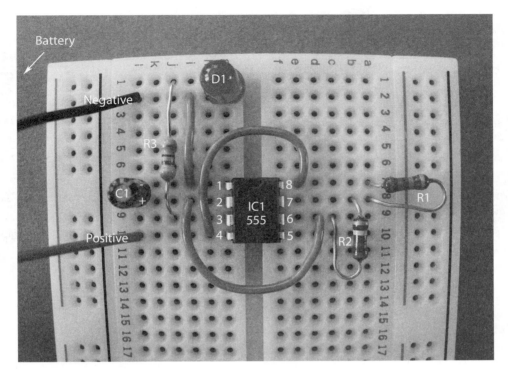

FIGURE 5-4 The breadboard layout for the LED flasher

Time to Experiment!

Once you are happy with your breadboard layout, connect the 4.5 volt battery to the battery clip; the LED should flash on and off at a steady rate. Your finished experimental circuit layout should look like Figure 5-7.

NOTE

Refer to Chapter 3 for building breadboard layouts and fault-finding guidelines.

Notice in Figure 5-4 how the negative lead of the battery is connected to the flat side of the LED and the negative side of capacitor C1.

BE CAREFUL!

It is important that you connect the electrolytic capacitor (C1) the correct way round in the circuit; if you don't, it could leak or explode!

Figures 5-5 and 5-6 show some close-up views of the breadboard layout, taken at different angles, which will help you build the circuit.

FIGURE 5-5 Notice the position of the semicircle of IC1 and the flat side of the LED.

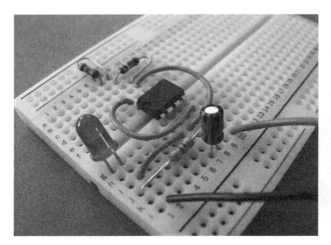

FIGURE 5-6 See the markings on capacitor C1, which identify the negative and positive leads.

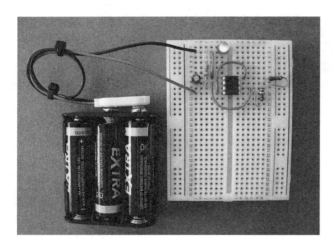

FIGURE 5-7 Connecting the 4.5 volt battery will cause the LED to flash on and off like a mobile phone message indicator.

If your circuit works, the LED will flash on and off at a steady rate until the batteries are removed or they run dry. If the LED does not flash on and off, you need to remove the battery immediately and check that your breadboard layout matches the layout shown in the figures. Also make sure that IC1, the capacitor C1, and the LED D1 are fitted in the correct orientation; otherwise, the circuit will not work correctly.

You can calculate the astable timings for this circuit by substituting the component values into the formulas described earlier in the chapter.

How to Convert Capacitor and Resistor Values into Numbers That You Can Use in the Formulas

You need to convert the capacitor value to farads and the resistor value to ohms when using the astable formulas. You can use the following calculations to help you do this.

Capacitance

- To convert picofarads (pF) to farads, *divide* the pF value by 1,000,000,000,000.
- To convert nanofarads (nF) to farads, *divide* the nF value by 1,000,000,000.
- To convert microfarads (µF) to farads, *divide* the µF value by 1,000,000.

Resistance

- To convert kilo-ohms (KΩ) to ohms (Ω), *multiply* the KΩ value by 1000.
- To convert mega-ohms (KΩ) to ohms (Ω), *multiply* the MΩ value by 1,000,000.

Here's how to calculate the LED on time:

$$On = 0.693 \times (R1 + R2) \times C1$$
$$On = 0.693 \times (22K\Omega + 220K\Omega) \times 2.2\mu F$$
$$On = 0.693 \times (22,000\Omega + 220,000\Omega) \times 0.0000022F$$

LED On time = 0.37 seconds

Here's how to calculate the LED off time:

$$Off = 0.693 \times R2 \times C1$$
$$Off = 0.693 \times 220K\Omega \times 2.2\mu F$$
$$Off = 0.693 \times 220,000\Omega \times 0.0000022F$$

LED Off time = 0.34 seconds

Here's how to calculate the frequency of the LED flash. First, the total time period for one cycle:

$$T = T1 + T2$$
$$T = 0.37 + 0.34 = 0.71 \text{ seconds}$$

Then you can calculate the frequency of oscillation:

$$F = 1 / T$$
$$F = 1 / 0.71$$
$$\text{The LED flashing frequency} = 1.4 \text{ Hz}$$

This means that the on/off cycle of the LED occurs roughly one-and-a-half times every second.

TABLE 5-1 Component Table for Useful Astable Timings

R1	R2	C1	LED On (secs)	LED Off (secs)	Total Time (secs)	Frequency (Hz)	Notes
22KΩ	220KΩ	2.2µF	0.37	0.34	0.71	1.4	Message indicator: The LED flashes at a regular rate.
220KΩ	220KΩ	2.2µF	0.67	0.34	1.01	0.99	Red alert: The LED on time is longer than the off time.
18KΩ	18KΩ	2.2µF	0.05	0.03	0.08	12.2	Strobe light: The LED flashes at a very fast rate.

Further Experimentation

Try replacing the two resistors R1 and R2 with different values to see what effect this has on the flash rate of the LED. Make sure that you remove the battery from the breadboard before making any changes to these component values.

BE CAREFUL!

Do not alter the value of resistor R3, because this could damage the LED.

You could also try altering the value of the capacitor (C1) to see what effect this has on the LED flash rate. If you use an electrolytic capacitor, make sure that you connect it the correct way around in the circuit so that the negative lead is connected to pin 1 of IC1.

HINT!

If you alter the component values and the LED looks as though it is not flashing, try to calculate on and off timings using the formulas. You might find that the LED is actually flashing at such a fast rate that it looks like the LED is permanently illuminated to the human eye!

Table 5-1 shows a few example component values to get you started. The table also shows some blank spaces so that you can write down some of the component values that you find useful in your experiments.

Summary

In this chapter, you learned how to use a 555 timer IC to flash an LED on and off. Can you think of any possible uses for this circuit? Can you think of electronic products that you have seen that use this type of flashing LED effect? You will see this circuit again in later experiments, which might give you some other ideas.

CHAPTER 6

How Do Screens Show Colors? Experiment with an RGB LED

IF YOU HAVE BEEN FOLLOWING the experiments in sequence, you have already learned how to illuminate an LED and to calculate its series resistor value. You will have also discovered how to create a 555 astable timer circuit that can be used to flash an LED on and off. Experiment 3 in this chapter explores how a color screen on a mobile phone is able to create different colors. You'll be introduced to a different type of LED that contains three different colors in a single component.

How to Create Colors

Before you build the electronic circuit for this project, you'll find it helpful to understand how the human eye visualizes color. White light actually comprises a range of colors that, when combined, look white to the naked eye. You can see the range of colors in sunlight when you look at a rainbow. The colors that you see on a mobile phone screen, or in fact any type of color screen such as your TV or computer screen, use a similar method to create colors. By mixing just three light colors to create an image—red, green, and blue—a color screen can show multitudes of different colors. Look at the diagram in Figure 6-1, which shows how some colors can be created by using only red, green,

and blue. So, for example, if you mix blue and red together, you can create a violet color.

Figure 6-2 shows a photo of a calculator that was taken on a mobile phone.

If you place a magnifying glass or microscope over the mobile phone screen, you will see tiny dots (called *pixels*) that make up the image on the screen. Each of the screen pixels consists of three different color elements—red, green, and blue. Figure 6-3 shows a close-up of the % button on the calculator photograph, which shows the individual pixels in the image.

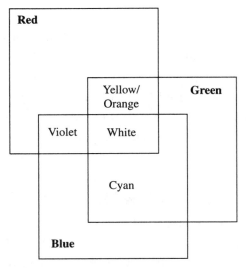

FIGURE 6-1 Red, green, and blue can be merged together to create other colors.

FIGURE 6-2 A photograph of a calculator shown on a mobile phone screen

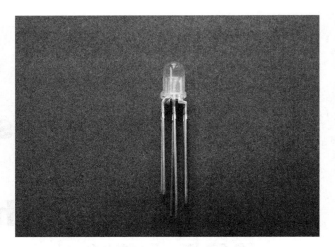

FIGURE 6-4 A typical RGB LED

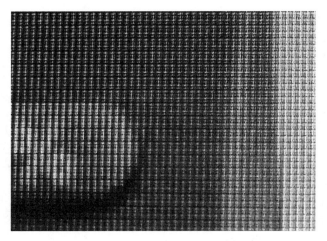

FIGURE 6-3 You can see individual pixels when you zoom in.

In this experiment, you will emulate one of these individual pixels to create various colors using a multicolor LED. Although a mobile phone screen doesn't always use LED technology to create the image, this experiment uses the same color principles to demonstrate how you can create different colors using an LED.

Red, Green, and Blue LED

So far, you've used a red LED in your experiments; this experiment uses a red, green, and blue (RGB) LED that contains these three color elements inside a single package. Figure 6-4 shows a typical RGB LED.

A normal LED has two leads (an anode and a cathode) that allow you to connect it to the circuit, as explained in Chapter 4. Because an RGB LED contains three colors, it normally has six connections—two leads for each LED color. Figure 6-5 shows the pin layout for the RGB LED used in this experiment. You can see that this LED actually contains two blue elements but you will only be using one in this experiment. If you use a different type of RGB LED, it may have slightly different lead connections, and you can check these out on the relevant manufacturer or supplier's datasheet.

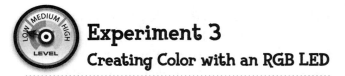

Experiment 3
Creating Color with an RGB LED

This experiment shows you how you can illuminate an RGB LED in various ways to create the various colors shown in Figure 6-1. You will also experiment with various resistor values to create some other interesting color combinations.

The Circuit Diagram

The circuit diagram for the experiment is shown in Figure 6-6.

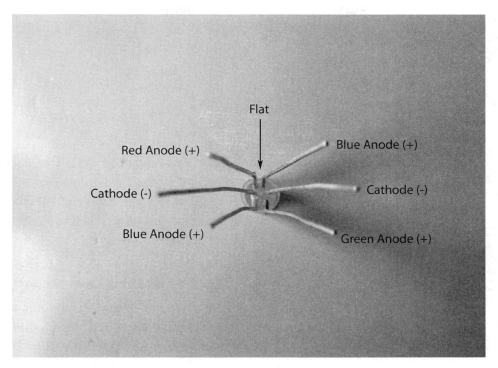

FIGURE 6-5 The pin layout for the RGB LED used in this experiment

How the Circuit Works

This circuit is constructed using three building blocks: power supply, output, and input. The circuit is powered by three 1.5 volt AA batteries wired in series to create a 4.5 volt supply. The RGB LED (D1) and the three resistors (R1, R2, and R3) create the output part of this circuit. This part of the circuit is very straightforward and is very similar to the LED experiment in Chapter 4, but this time three series resistors are included—one for each LED color. If you look at the circuit symbol for the RGB LED (D1), you will see that it contains three individual LEDs, one for each color.

The three switches SW1 to SW3 create the input part of the circuit. You are not going to use actual switches in this experiment; instead, you will be using some wire links. When any of the wire link switches is connected between the positive lead of the battery and one of the three resistors (R1 to R3), it allows current to flow into each individual color LED, and you will soon see what effect this has on the LED color.

LED Resistor Calculations

From Chapter 4, you know that the formula for calculating an LED series resistor is as follows:

$$R = (V - V_F) / I_F$$

- **R** is the value of the series resistor (measured in ohms, Ω)
- **V** is the voltage of the circuit (measured in volts)
- **V$_F$** is the typical forward voltage drop of the LED (measured in volts)
- **I$_F$** is the amount of current that flows through the LED (measured in amps)

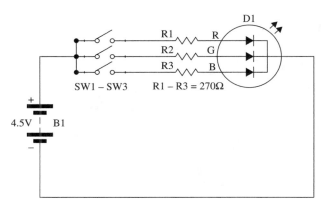

FIGURE 6-6 The circuit diagram for the RGB LED experiment

The V ratings of a green and blue LED are usually different from that of a red LED, so the series resistor for each color has to be calculated individually. We need to make sure that the maximum current that flows through each color LED is no more than 25mA, and with fresh batteries, the voltage could be nearly 5 volts. The calculations for this experiment are shown next.

For the red LED:

$$R = (V - V_F) / I_F$$
$$V_F \text{ (typical)} = 2.0V$$
$$R = (V - V_F) / I_F$$
$$R = (5V - 2.0V) / 0.025A$$
$$R = 3V / 0.025A$$
$$R = 120\Omega$$

For the green LED:

$$R = (V - V_F) / I_F$$
$$V_F \text{ (typical)} = 2.2V$$
$$R = (V - V_F) / I_F$$
$$R = (5V - 2.2V) / 0.025A$$
$$R = 2.8V / 0.025A$$
$$R = 112\Omega$$

For the blue LED:

$$R = (V - V_F) / I_F$$
$$V_F \text{ (typical)} = 4V$$
$$R = (V - V_F) / I_F$$
$$R = (5V - 4V) / 0.025A$$
$$R = 1V / 0.025A$$
$$R = 40\Omega$$

Ohm's Law states that as resistance increases the current flow decreases, so you can see that if you use a 270Ω resistor for each LED color, this will reduce the current well below the 25mA maximum calculated.

The Breadboard Layout

Using your breadboard and the components in the parts list, build the circuit layout, which is shown in Figure 6-7. Another close-up, taken at a different angle, is shown in Figure 6-8. You will notice that the negative of the battery is also connected to the right-hand cathode connection (−) of the LED that I used; this is to ensure that the red element of the LED will illuminate properly.

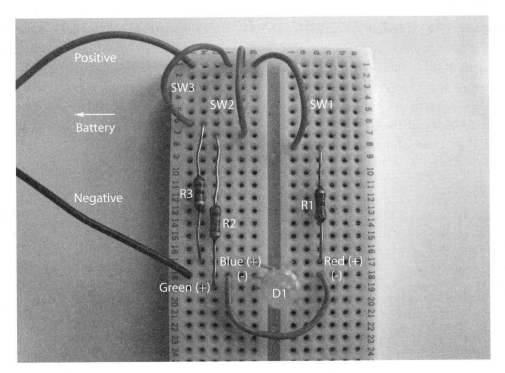

FIGURE 6-7 The layout for the RGB LED experiment

Things You'll Need

The components and equipment that you will need for this experiment are outlined in the following table. Locate and prepare the items that you need before starting the experiment.

Code	Quantity	Description	Appendix Code
D1	1	5mm RGB LED	6
R1	1	270Ω 0.5W ±5% tolerance carbon film resistor	—
R2	1	270Ω 0.5W ±5% tolerance carbon film resistor	—
R3	1	270Ω 0.5W ±5% tolerance carbon film resistor	—
B1	1	4.5V battery holder	15
B1	3	1.5V AA batteries	—
B1	1	PP3 battery clip	17
—	1	Breadboard	1
SW1, SW2, SW3	—	Wire links	—
R1, R2, R3*	—	Various 0.5W ±5% tolerance carbon film resistors with values greater than 180Ω	—

*See the "Further Experimentation" instructions at the end of this chapter.

NOTE

The Appendix Code column of the table refers to specific parts that I used in this experiment. Information about sourcing these parts is outlined in the Appendix.

NOTE

Refer to Chapter 3 for building breadboard layouts and fault-finding guidelines.

Time to Experiment!

Once you have built the breadboard layout, connect the 4.5 volt battery supply to the board, as shown in Figure 6-9, making sure that each of the wire links (SW1 to SW3) is connected to the positive battery

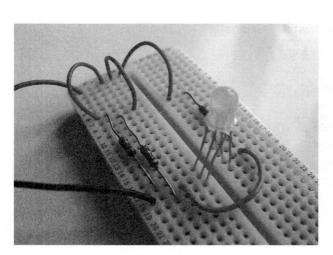

FIGURE 6-8 Close-up of the RGB LED layout

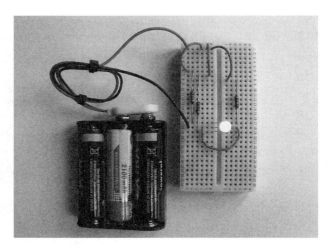

FIGURE 6-9 Power up the circuit with all three switches activated.

connection. The LED should illuminate with a blue/white color.

If the LED doesn't illuminate, or if it is a different color, remove the battery immediately and check that the LED leads are connected to the resistors as shown in the figures.

Mixing Colors

Now for the fun part: Let's make some colors. The following individual experiments show you how to create the various colors shown in the diagram in Figure 6-1.

Start by removing all of the wire links. Then connect SW3 to the positive battery rail, which is any of the holes in the top left-hand row 1 of the breadboard (holes 1G to 1L in Figure 6-10) and leave the SW1 and SW2 links unconnected. Then add the battery, and you should see that the LED shines with a blue color, as shown in Figure 6-10.

HINT!

Try this experiment in a darkened room to make it easier to see the color.

Next, disconnect wire link 3 and connect wire link 2 to the positive battery connection instead. The LED lights a green color.

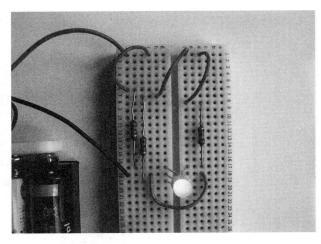

FIGURE 6-10 SW3 activated makes the LED light blue. Notice that when not in use the two other wire links are simply connected to row 1A to 1F of the breadboard, which has no voltage applied to it.

Finally, try various wire link configurations to see some of the different colors that the RGB LED can create.

Table 6-1 shows the various wire link switch configurations to make various colors. Use the spaces in the table to write down the colors that you create.

NOTE

"On" in the table means that the wire link is connected to the positive battery connection. "Off" means that the wire link is not connected.

TABLE 6-1 LED Colors Created by Changing the Various Switch Settings

SW1 and R1 Value	SW2 and R2 Value	SW3 and R3 Value	LED Color
Off	Off	Off	
Off	Off	On (270Ω)	
Off	On (270Ω)	Off	
Off	On (270Ω)	On (270Ω)	
On (270Ω)	Off	Off	
On (270Ω)	Off	On (270Ω)	
On (270Ω)	On (270Ω)	Off	
On (270Ω)	On (270Ω)	On (270Ω)	

Further Experimentation

After you've had a chance to play around with the various switch settings, try changing some or all of the resistors to different values and try the various wire link settings to see what effect it has on the colors that you create. You may find that the color changes are very subtle, depending on the values of the resistors that you use. You can use the spaces in Table 6-2 to write down the results of your experimentation. Two examples are shown in the table to get you started.

BE CAREFUL!

Make sure that you use resistors with values that are greater than 120Ω; otherwise, you might damage the LED.

Summary

You have discovered that by altering the resistor values, you can either brighten or darken the color of each individual LED color element. When various colors are mixed together, they change

TABLE 6-2 Table to Complete Your Tests

SW1 and R1 Value (Red)	SW2 and R2 Value (Green)	SW3 and R3 Value (Blue)	LED Color
On (120Ω)	Off	On (2.7KΩ)	Pink
On (390Ω)	On (120Ω)	Off	Lime green
Off	On (Ω)	On (Ω)	
Off	On (Ω)	On (Ω)	
Off	On (Ω)	On (Ω)	
Off	On (Ω)	On (Ω)	
On (Ω)	Off	On (Ω)	
On (Ω)	Off	On (Ω)	
On (Ω)	Off	On (Ω)	
On (Ω)	Off	On (Ω)	
On (Ω)	On (Ω)	Off	
On (Ω)	On (Ω)	Off	
On (Ω)	On (Ω)	Off	
On (Ω)	On (Ω)	Off	
On (Ω)	On (Ω)	On (Ω)	
On (Ω)	On (Ω)	On (Ω)	
On (Ω)	On (Ω)	On (Ω)	
On (Ω)	On (Ω)	On (Ω)	

the color of the LED. This is similar to how the individual pixels on a mobile phone screen are able to create various colors to build a complex image. In reality, the phone screen doesn't use various value resistor values to do this; instead, it uses complex electronic circuitry to control the brightness of each individual pixel.

This experiment should also make you appreciate how small electronics can be. The LED that you used in the experiment is only 5mm in diameter, and a mobile phone screen might have a screen resolution of 640 × 480 pixels, which means that the screen actually contains 307,200 individual pixels. If you were to build a screen using 5mm LEDs, the mobile phone screen would have to be 3.2 meters wide × 2.4 meters high! That's one big phone!

CHAPTER 7

Can We Talk? Build a Working Telephone

ONE VERY IMPORTANT FEATURE of a mobile telephone is that it allows you to talk to someone! If you have read Chapter 4, you will realize that a mobile phone contains an earpiece, which allows you to listen to the other person, and a microphone, which allows your voice to be converted into an electronic signal that can be heard by the person on the other end of the phone.

This chapter's Experiment 4 shows you how to create a basic telephone by using a few components and a microphone and earpiece. Two of these circuits are then combined on a single piece of breadboard so that you can have a two-way conversation with someone in another room.

Basic Telephone Setup

Before you start this experiment, you'll find it helpful to understand how a basic telephone connection can be created. This experiment doesn't show you how to create a wireless mobile phone signal to communicate across a cellular network, but

it does show you how to create a hardwired circuit that mimics the operation of a landline telephone.

The block diagram in Figure 7-1 shows you how a basic telephone circuit operates.

Figure 7-1 shows that a telephone circuit requires a four-core cable to link the two telephone receivers together. This connection could be made in a number of ways. The block diagram shows that in this example the microphone from circuit 1 is positioned next to the earphone from circuit 2 to create the remote telephone receiver.

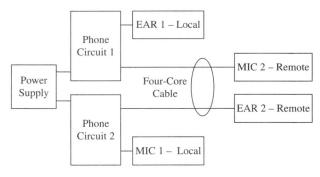

FIGURE 7-1 A basic telephone setup

INTERESTING FACT

Scottish inventor Alexander Graham Bell (1847–1922) designed and built the first ever working telephone back in 1876 when he was only 29 years old.

Experiment 4
Building a Working Telephone

This experiment shows you how the building blocks shown in Figure 7-1 can be brought to life by using just a handful of electronic components. You will initially construct half of a telephone (a single microphone and earpiece), and then go on to adding the second part of the circuit to create a working telephone, which will allow you to have a two-way conversation with a friend.

The Circuit Diagram

The circuit diagram for the experimental telephone in Figure 7-2 shows a single transmitter/receiver circuit. You will eventually need to build two

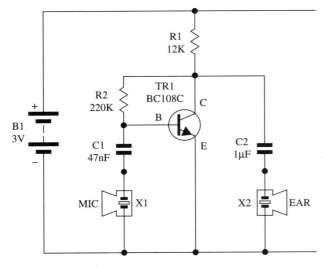

FIGURE 7-2 The circuit diagram for one-half of a telephone transmitter/receiver

of these circuits to create a complete two-way telephone.

How the Circuit Works

This circuit is classed as an *analog* circuit, because it converts sound that enters the microphone into a moving electrical signal waveform, which is then converted back into a sound signal when it enters the earpiece. This circuit is constructed using four building blocks: power supply, input, control circuitry (amplifier), and output.

This circuit is powered by two 1.5 volt AA batteries (B1) wired in series to create a 3 volt supply. The microphone (X1) is actually a crystal earpiece, which works well as a microphone in this experiment.

TR1, R1, R2, C1, and C2 form the control circuitry. The circuit is a basic amplifier circuit that is able to run from the 3 volt supply. The heart of the amplifier is an NPN transistor (you read about transistors in Chapter 3).

Here's how the circuit works: The base of the transistor receives the input voltage signal from the microphone, and it converts this into a larger voltage output that is fed across the collector and emitter of the transistor. This changing voltage signal is mimicked in the earpiece, which is wired across the collector and emitter, and converts this into a reproduction of the sound heard through the microphone.

The two capacitors (C1 and C2) help to filter out any unwanted voltage signals to provide a clearer sound signal. The values of the two resistors, R1 and R2, provide a good level of amplification and sound level in this experiment.

The earpiece (X2), which is a second crystal earpiece, converts the amplified electrical signal from the amplifier control circuitry into sound that you can hear.

Things You'll Need

The components and equipment that you will need for this experiment are outlined in the following table. Locate and prepare the items that you need before starting the experiment.

Code	Quantity	Description	Appendix Code
TR1-2	2	BC108C NPN transistor	9
R1/R3	2	12KΩ 0.5W ±5% tolerance carbon film resistor	—
R2/R4	2	220KΩ 0.5W ±5% tolerance carbon film resistor	—
C1/C3	2	47nF ceramic disc capacitor	—
C2/C4	2	1µF boxed polyester capacitor	—
X1–X4	4	Crystal earpiece	21
B1	1	3V battery holder	14
B1	2	1.5V AA batteries	—
B1	1	PP3 battery clip	17
—	1	Breadboard	1
—	—	Wire links	—
—	1	12-way terminal block	—
—	1	Four-core telephone cable	—

NOTE

The Appendix Code column of the table refers to specific parts that I used in this experiment. Information about sourcing these parts is outlined in the Appendix.

Crystal Earpiece

This circuit uses four crystal earpieces (X1–X4), which have not been discussed in the preceding chapters. They look similar to the type shown in Figure 7-3.

A crystal earpiece fits into your ear and contains a piece of crystal connected to wires. Whenever a moving signal is presented across the wires, the crystal vibrates, and this converts the electrical signal into a sound signal that you can hear. (It is quite a sensitive device and can be used in an unconventional way; it can also be used as a microphone—even though it is not designed to do this, it works! You'll see how as you start to build the experiment.) The crystal earpiece is not polarity sensitive, so it doesn't matter which way around you connect the two wires in the circuit; in this circuit, however, it works better in one direction, so you might need to switch the wires around to find the best mode of operation.

FIGURE 7-3 The crystal earpiece on the right does not include the bit that fits in your ear so that you can see the crystal inside. This also makes it easier to talk into.

The Breadboard Layout

Using your breadboard and the components in the parts list, build the circuit layout shown in Figure 7-4. You'll notice that at this stage, you'll have some components left over. Save these for the next stage of the experiment, which is described shortly.

NOTE

Refer to Chapter 3 for building breadboard layouts and fault-finding guidelines.

Figure 7-5 shows a close-up of the breadboard layout at a different angle.

The final layout with the microphone and earpiece fitted is shown in Figure 7-6.

NOTE

You can read more about the transistor used in this experiment and its pin connections in Chapter 3.

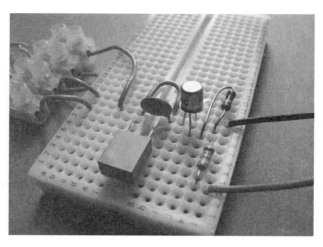

FIGURE 7-5 Close-up of the telephone transmitter/ receiver

Time to Experiment!

After you have built the breadboard layout, connect the 3 volt battery holder to the PP3 battery clip and place the earpiece (X2) into your ear. Now pick up the microphone (X1) and move it around in your hand. You should hear a rustling noise in the earpiece; this is the noise that the microphone is transmitting to the earpiece. If you hear the rustling, hold the microphone in your

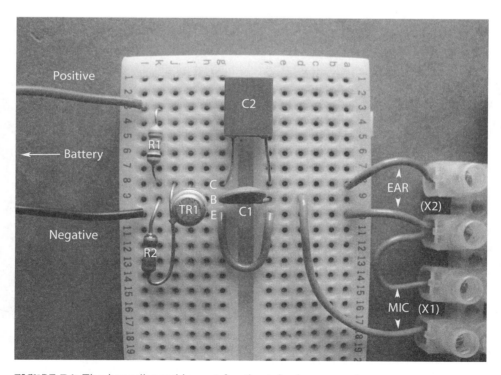

FIGURE 7-4 The breadboard layout for the telephone receiver

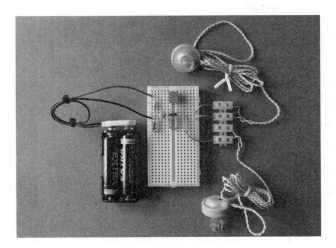

FIGURE 7-6 The completed layout

hand and talk into it; you should hear your voice clearly in the earpiece!

If it doesn't work, try switching around the two cables of the earpiece. If the circuit still doesn't work, do the same with the microphone; this can make a difference regarding how clearly you can hear the signal.

If you want, you could use a 20-foot-long piece of two-core cable and wire the microphone in a separate room, and then ask a friend to talk into it while you listen with the earpiece. You should be able to hear the other person clearly. The only problem with this setup, however, is that you can hear your friend, but he or she will be unable to hear you talking—your friend doesn't have an earpiece, and you don't have a microphone. So you will need to use the spare components that you saved to build an identical circuit. The complete circuit diagram is shown in Figure 7-7.

You can build this second circuit on the breadboard and wire it all together to create a layout like the one shown in Figure 7-8. This layout is similar to the block diagram in Figure 7-1.

NOTE

Notice how the positive and negative leads of the battery are connected to wire links that provide power to both halves of the circuit.

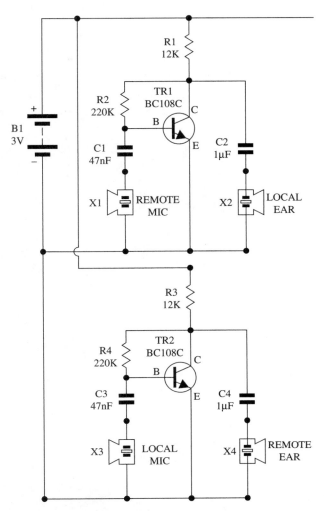

FIGURE 7-7 The complete circuit diagram for the two-way telephone

The local microphone and earpiece shown in Figure 7-8 is your telephone. Now, using a four-way terminal block and a four-core cable, you can connect a remote microphone and earpiece to a matching four-way terminal block in another room; this is your friend's telephone. This second receiver setup is shown in Figure 7-9.

If your friend talks into a microphone you should be able to hear his or her voice clearly, and if you speak into your microphone, your friend should be able to hear you in his or her earpiece. Congratulations! You have built a working telephone!

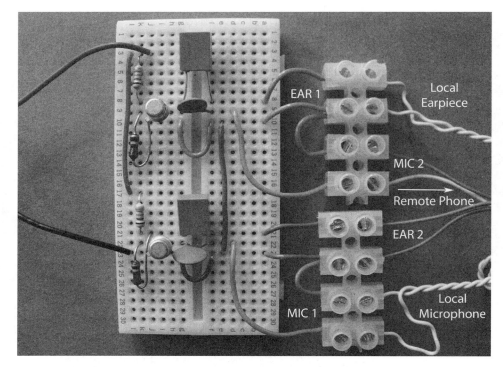

FIGURE 7-8 Build the second half of the circuit on the breadboard.

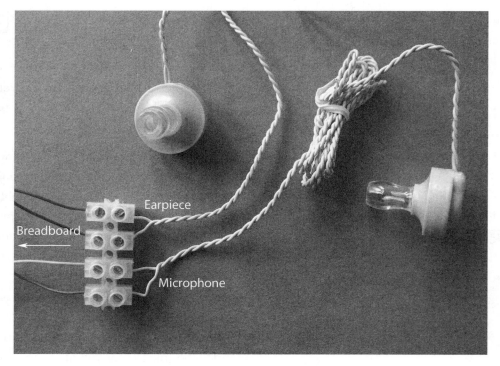

FIGURE 7-9 A second telephone receiver for your friend

The circuit is very efficient, drawing only around 0.5mA, and this means that the two AA batteries will last a number of weeks before they have to be replaced.

HINT!

Always make sure to remove the batteries from your circuit when it is not in use.

Further Experimentation

This circuit is very flexible and will work with a wide range of component values, although the sound quality of the telephone will vary depending on the component values that are used. If you are unable to find a BC108C transistor, you could use an alternative NPN transistor that has a fairly high gain. I tested a BC267B NPN transistor and this seemed to work fine in this circuit. Some transistors will not work well, however, such as the BC109.

Try altering the capacitor values to see what effect this has on the quality of the sound. Reducing the value of R2 also reduces the volume and tone of the sound that you hear in the earpiece.

Summary

In this chapter you learned how to create an amplifier circuit using a transistor and a couple of resistors and capacitors. This circuit enabled the sound of human voice to be converted into an electrical signal via a microphone. This electrical signal was then amplified (made larger) and converted back into an electrical signal, which was large enough so that you could hear it in an earpiece. You also discovered that by combining two identical amplifier circuits together you can create a single circuit that emulates a real telephone.

PART THREE

Behind the Wheel: Dashboard Experiments

CHAPTER 8

Which Way Now? Create Indicator Lights

THIS SECTION OF THE BOOK IS DEDICATED TO a collection of experiments that attempt to emulate some of the features that you might find in a car. Before we build the first dashboard experiment, think about some of the features that you might find on a car's dashboard.

Features of a Typical Car Dashboard

Look at Figure 8-1 and Table 8-1, which show and describe some of the features that you might find associated with a car's dashboard.

FIGURE 8-1 A typical car dashboard

In this chapter, you'll build some experiments that relate to some of the features shown in Table 8-1.

Experiment 5
Creating Indicator Lights

In this experiment, you'll create indicator lights that emulate turn signals on a car by using two LEDs as the left and right turn signal lights on a dashboard. You've probably noticed that when the turn signals in a car are activated, they make a clicking noise in time with the flashing lights. This is because the lights use a *relay* to switch the indicators, which makes this noise. Our circuit will also include a relay, which means that it will make the same type of clicking noise.

BE CAREFUL!

This experiment shows you how to flash an LED. If you suffer from epilepsy or are affected by flashing lights, this experiment is not for you.

What Is a Relay?

Before you take a look at the circuit and its operation, you should understand what a relay

TABLE 8-1 Key Features of a Typical Car Dashboard

Feature	Description
Indicator lights/turn signals*	Show other drivers which direction you intend to travel
Windshield wiper control	A button or switch activates wipers that remove rain from the windshield
Horn*	A part of the steering wheel activates a loud noise to warn other people on the road
Temperature sensor*	Measures the temperature of the engine
Speedometer/odometer	Shows how fast you are traveling and the amount of miles that you have traveled
RPM meter	Shows how many revolutions per minute the engine is making
Fuel Gauge	Shows you how much fuel is inside the petrol tank
Oil light	Illuminates if the engine needs more oil
Rain sensor*	Senses, outside of the car, when it is raining to activate the windshield wipers automatically

*The features that we are going to explore are identified by an asterisk.

does and how it works. Relays are electronic switches that allow you to use a small current and voltage to switch a much larger current and voltage without the two different circuits mixing together. It does this by using an *electromagnet* to activate the switch. The electromagnet is created when a voltage is applied to the *coil* of the relay (see Figure 8-2). So, for example, in a car, the indicator lights will be drawing a fairly large current when they are switched on and off, compared to the control input current to the relay.

Here's the circuit symbol for a typical double-pole, double-throw relay:

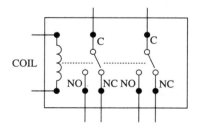

The "curly" connections on the symbol represent the coil of a relay, and the other connections are the switch contacts. A typical relay is shown in Figure 8-2; the electromagnetic coil is on the left, and the switch contacts are on the right.

Following is some terminology you'll come across when reading about or purchasing relays:

- **Coil** The coil will have a voltage rating, and you need to make sure that this is the same or slightly higher than the voltage of your control circuit.

- **Switch contacts** Contacts also have voltage and current ratings, which need to be higher than the voltage and current rating of your output circuit.

FIGURE 8-2 Inside a typical relay

- **Normally open (NO)**   This relates to the relay contacts; it means that when the relay is not energized, these contacts are not touching and are in an open position, creating an open circuit.

- **Normally closed (NC)**   This relates to the relay contacts; it means that when the relay is not energized, these contacts are touching and are in a closed position, creating a short circuit.

- **Single pole**   This relay will contain one set of switch contacts, a common connection, and a normally open and normally closed set of contacts.

- **Double pole**   This relay will contain two sets of common, normally open and normally closed contacts.

NOTE

Details of other electronic components are explained in Chapter 3.

The Circuit Diagram

Now let's take a look at the circuit diagram for the indicator lights shown in Figure 8-3.

How the Circuit Works

This circuit is made up of four building blocks: power supply, input, control circuitry (555 astable pulse generator), and an output. This circuit is powered by four 1.5 volt AA batteries that are wired in series to create a 6 volt power supply. The "turn" switch (SW1) is the input for this circuit and makes one of the indicator LEDs illuminate. This control circuitry should be familiar to you if you have already read Chapter 5, which showed you how to make a flashing LED. The control circuit contains a 555 timer (IC1) configured in astable mode, and its on-off time is set by the values of R1, R2, and C1. The two LEDs (D1 and D2) and their series resistors (R3 and R4), in conjunction with the relay circuitry (RL1, D3 and D4), create the output part of this circuit.

Here's how the circuit works: The 555 timer, which is set up in astable mode, creates a regular pulse output from pin 3 of this device. This pulsed signal activates the coil of the relay (RL1) and switches the relay on and off at a regular interval. The purpose of diode D4 is to make sure that only the positive signal from the 555 timer output reaches the coil of the relay. D3 acts as a *flywheel*

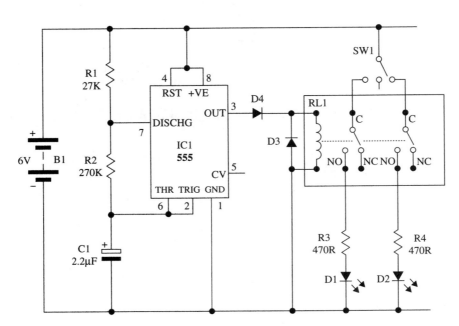

FIGURE 8-3 The circuit diagram for the indicator lights

(or flyback) diode, and it is important that this is included, because when the coil deactivates, a small current feeds out from the coil connections as the electromagnetic coil collapses. This small current could damage the 555 timer. Because diode D3 is included, it makes the current flow in a circle back through the coil until it eventually disappears; this stops the collapsing current reaching the 555 timer.

INTERESTING FACT

A flyback diode is used to eliminate flyback, which is the sudden voltage spike experienced across an inductive load when its supply voltage is suddenly reduced or removed.

The Breadboard Layout

Using the components in the parts list, build the circuit by following the breadboard layout for the indicator lights, as shown in Figure 8-4.

Figures 8-5 and 8-6 show close-up images of the breadboard layout that will help you to build the circuit.

Time to Experiment!

Once you have built the circuit, connect the wire link that is acting as the switch (SW1) between the positive breadboard rail to the common connection of the left side of the relay, and then connect the battery to the circuit, as shown in Figure 8-7.

Things You'll Need

The components and equipment that you will need for this experiment are outlined in the following table. Prepare the items that you need before starting the experiment.

Code	Quantity	Description	Appendix Code
IC1	1	555 timer	18
R1	1	27KΩ 0.5W ±5% tolerance carbon film resistor	—
R2	1	270KΩ 0.5W ±5% tolerance carbon film resistor	—
C1	1	2.2µF electrolytic capacitor (minimum 10 volt rated)	—
D1/D2	2	5mm red LED	4
R3/R4	2	470Ω 0.5W ±5% tolerance carbon film resistor	—
D3/D4	2	1N4003 diode	—
RL1	1	Double-pole, double-throw relay with a 6 volt DC coil	23
SW1	1	Wire link	—
B1	1	6V battery holder	16
B1	4	1.5V AA batteries	—
B1	1	PP3 battery clip	17
—	1	Breadboard	2
—	—	Wire links	—

NOTE

The Appendix Code column of the table refers to specific parts that I used in this experiment. Information about sourcing these parts is outlined in the Appendix.

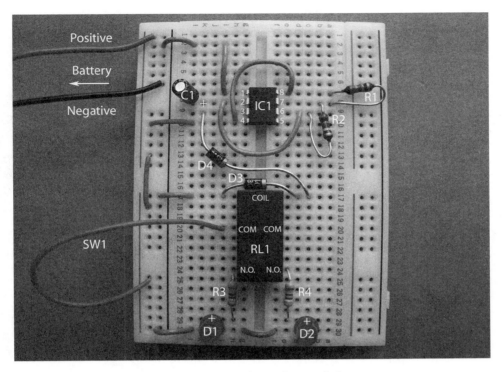

FIGURE 8-4 The breadboard layout for the indicator lights

You should find that the relay starts to click and the left-hand LED starts to flash on and off in time with the clicking noise, just like real turn signals in a car.

Now try moving the wire link from the left side of the relay to the right side. Figure 8-8 shows that this time the right-hand LED flashes on and off in time with the relay.

Further Experimentation

Try a couple more experiments after you have finished playing with the indicator lights.

- See if you can modify the circuit so that *both* LEDs flash on and off in time with the relay, just like the hazard lights on a car.

- Once you have figured that one out, see if you can modify the circuit to make the LEDs flash on and off alternately, so that when the left LED flashes on, the right LED flashes off, and then when the left LED flashes off, the right LED flashes on.

 HINT!

Don't forget that the relay contains both normally open and normally closed contacts.

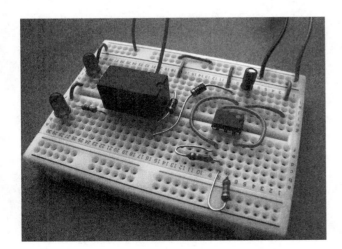

FIGURE 8-5 Close-up of the breadboard layout

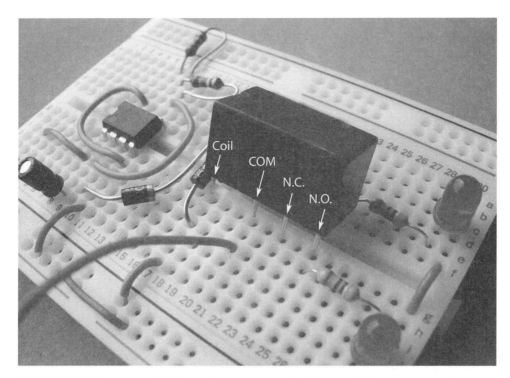

FIGURE 8-6 Close-up with the relay removed slightly so that you can see the pin connections. Notice that the flat sides of the two LEDs are positioned toward the bottom of the board.

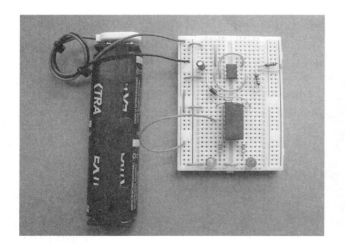

FIGURE 8-7 "Turning left"

FIGURE 8-8 "Turning right"

Summary

In this chapter, you learned how to make a basic 555 timer circuit drive a relay, which could be used to drive LEDs or indicator lights with a higher current rating than the 555 timer can handle. You also discovered how a relay operates and how diodes can be used in a circuit to protect sensitive electronic circuitry from "flyback" from a relay coil.

CHAPTER 9

Is It Hot in Here? Build a Temperature Sensor

MOST CAR DASHBOARDS HAVE A TEMPERATURE gauge that shows the relative temperature of the engine. This readout warns you if the engine is getting too hot, because an overheated engine can mean that something might be wrong and you should stop the car. Most dashboards also have an indicator light that illuminates when the engine reaches a temperature that is too high; you probably won't be able to see it on the dashboard because it is normally switched off. It might look something like Figure 9-1 when it's on.

In this experiment, you'll create a circuit that uses a temperature sensor that illuminates an LED if the temperature around the sensor starts to rise.

FIGURE 9-1 A typical temperature indicator on a dashboard

Of course, this circuit is not designed to work in a car engine, but you could use this circuit for some other applications, and some of these are suggested at the end of the chapter.

Experiment 6
Building a Temperature Sensor

Before you look at the circuit diagram for the temperature sensor experiment, you need to understand how the temperature sensor component used in this experiment operates.

What's a Thermistor?

In this experiment, you will discover a component that is part of the resistor family of components: a *thermistor*. The circuit symbol for a thermistor looks like this:

A thermistor has a resistance, just like a resistor, but the resistance of a thermistor changes depending on the temperature around it. You'll find several different types of thermistor available with different resistances at different temperatures. There are two main types:

- **Negative temperature coefficient (NTC)** The resistance of this type of thermistor reduces as the temperature around it increases.

- **Positive temperature coefficient (PTC)** The resistance of this type of thermistor increases as the temperature around it increases.

You will be using an NTC thermistor in this experiment, which is shown in Figure 9-2. This thermistor has a resistance of around 5KΩ when the ambient temperature around it is 20°C.

You can see what happens to the resistance of a thermistor if you plug its leads into a piece of breadboard and measure its resistance with your multimeter set to the resistance settings. Depending on the ambient temperature, you should find that the resistance of the thermistor is approximately 5KΩ. If you press your finger and thumb on the plastic case of the thermistor, the resistance value should start to reduce as your body temperature is transferred into the component. When you start to experiment, you'll use this change in resistance to make an LED illuminate when the temperature of the thermistor reaches a certain level.

NOTE

Details of other electronic components are explained in Chapter 3.

Potential Divider Calculations

You'll explore a neat circuit technique in your experiment, called a "potential divider." This type of circuit layout is shown in Figure 9-3.

This type of circuit allows you to generate a specific output voltage by altering the values of the resistors. You can use this formula to calculate the output voltage:

Output Voltage = Input Voltage × (R2 / (R1 + R2))

So, for example, suppose that R1 and R2 are both 4.7KΩ, and that the input voltage (the supply voltage) is 4.5 volts. If you add these values to the formula, you can calculate the output voltage as follows:

Output Voltage = 4.5V × (4.7KΩ / (4.7KΩ + 4.7KΩ))
Output Voltage = 4.5V × (4.7KΩ / 9.4KΩ)
Output Voltage = 4.5V × 0.5
Output Voltage = 2.25V

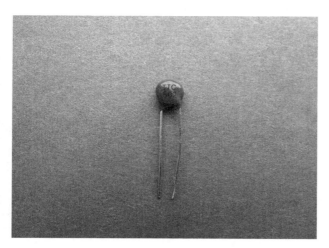

FIGURE 9-2 A thermistor

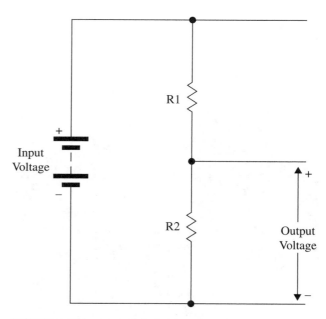

FIGURE 9-3 A potential divider circuit

So you can see from this calculation that the output voltage is half the input voltage when the values of R1 and R2 are equal.

Let's see what happens if you reduce the value of resistor R1 to 1KΩ and keep the value of R2 as 4.7KΩ.

Output Voltage = 4.5V × (4.7KΩ / (1KΩ + 4.7KΩ))
Output Voltage = 4.5V × (4.7KΩ / (5.7KΩ))
Output Voltage = 4.5V × 0.82
Output Voltage = 3.69V

This exercise shows you that you can increase the output voltage by reducing the resistance value of R1. Now that you understand this concept, you can put this circuit effect into good use in this experiment.

The Circuit Diagram

Have a look at the circuit diagram for the temperature sensor, shown in Figure 9-4. The circuit should now look familiar to you, because it uses the potential divider that you just read about.

How the Circuit Works

The temperature sensor circuit is made up of four building blocks: power supply, input, control circuitry, and output. This circuit is powered by three 1.5 volt AA batteries that are wired in series to create a 4.5 volt power supply. The thermistor (R1) is the input part of this circuit. Resistor R2, which is the static part of the potential divider, and the transistor (TR1) form the control part of the circuit. The LED (D1) and its series resistor (R3) are the output part of this circuit.

The circuit works as follows: In normal ambient temperatures of around 20°C (68°F), the resistance

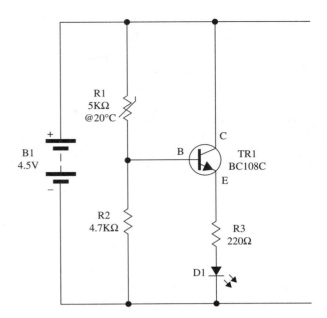

FIGURE 9-4 Circuit diagram for the temperature sensor

of the thermistor (R1) is around 5KΩ. This resistance, in conjunction with the 4.7KΩ resistor (R2), means that the base of the transistor has a voltage of half the supply voltage 2.25V—so the base is balanced, neither more positive than negative—so the transistor doesn't switch on and the LED is not illuminated.

As the temperature around the thermistor increases, the resistance of the thermistor is reduced. And as you saw earlier in the potential divider calculations, this increases the output voltage of the potential divider. So as the temperature increases, the voltage at the base of the transistor also increases and becomes more positive than negative. The transistor then starts to switch on and the LED illuminates. (You will be learning more about transistor switching circuits in Chapter 11.)

Things You'll Need

The components and equipment that you will need for this experiment are outlined in the following table. Prepare the items that you need before starting the experiment.

Code	Quantity	Description	Appendix Code
TR1	1	BC108C NPN transistor (or any high-gain transistor such as a BC267B)	9
R1	1	5KΩ @ 25°C NTC 500mW disc thermistor (TTC502)	11
R2	1	4.7KΩ 0.5W ±5% tolerance carbon film resistor	—
R3	1	220Ω 0.5W ±5% tolerance carbon film resistor	—
D1	1	5mm green LED	5
B1	1	4.5V battery holder	15
B1	3	1.5V AA batteries	—
B1	1	PP3 battery clip	17
—	1	Breadboard	1
—	—	Wire links	—

NOTE

The Appendix Code column of the table refers to specific parts that I used in this experiment. Information about sourcing these parts is outlined in the Appendix.

The Breadboard Layout

The breadboard layout for this experiment is very simple and is shown in Figure 9-5, which also shows the component numbers to refer to. The photograph in Figure 9-6 is taken at a slightly different angle so that you can see where the component pins are positioned.

HINT!

The pin connections for the transistor used in this experiment are shown in Chapter 3.

The breadboard layout is very simple and should take only a few minutes to build. Make sure that you carefully check your layout and then connect the battery to the circuit. You should find that the LED (D1) is not illuminated at this stage, or if it is, it will be very dimly lit.

Time to Experiment!

Now you have built the circuit and the battery is connected, touch the plastic casing (not the leads) of the thermistor gently using your finger and thumb, as shown in Figure 9-7.

After a few seconds, the LED will illuminate slightly. The heat from your fingers is transferred into the thermistor and starts to warm it up, and this starts to reduce the resistance of the thermistor, which activates the LED as described earlier. If you remove your finger and thumb from the thermistor, the LED should start to dim until it is no longer illuminated. You might need to try this experiment in a darkened room to see this effect.

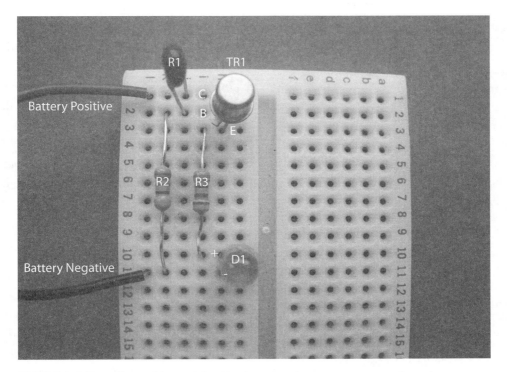

FIGURE 9-5 Breadboard layout for the temperature sensor experiment

To make the LED illuminate more brightly, you need to increase the temperature of the thermistor. Try using a hair dryer set on a medium heat setting to blow the warm air carefully over the thermistor for a few seconds, from a safe distance of around 2 feet away from the breadboard, as shown in Figure 9-8.

 BE CAREFUL!

Ask an adult to help you with this experiment. Also make sure that the heat from the hairdryer doesn't cause the breadboard and components to get too hot. Don't forget to switch off the hairdryer when you have finished with this part of the experiment.

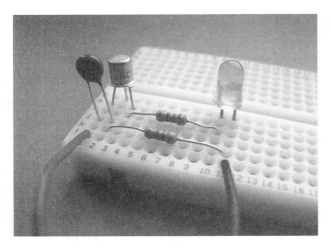

FIGURE 9-6 The breadboard from a different angle

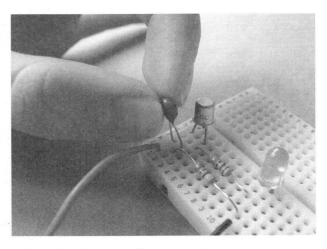

FIGURE 9-7 Use your finger and thumb to heat up the thermistor.

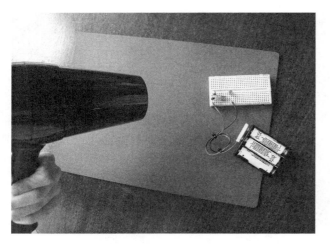

FIGURE 9-8 Blow warm air onto the thermistor.

As the temperature around the thermistor increases the LED will illuminate brightly. When this happens, switch the hairdryer off and watch to see how long it takes for the LED to switch off again.

 INTERESTING FACT

Electronic components don't like to be too hot, because they can't sweat like humans! If a component is worked too hard in a circuit it can become too hot, and a phenomenon called "thermal runaway" can occur, which causes the component to draw more and more current from the circuit as its temperature increases. This results in the destruction of the component when it finally exceeds its designed operating temperature.

Further Experimentation

Find yourself a calculator and try substituting different resistor values for R1 and R2 into the potential divider calculations that were shown earlier in the chapter. This will allow you to see what effect this has on the output voltage. These types of theoretical calculations are used by electronics designers and inventors so that they can understand how a circuit might behave before they start to build it.

Summary

In this experiment, you learned about the properties of a thermistor and what its circuit symbol looks like. You also learned about a potential divider circuit and how you can use the formulas to calculate what the output voltage of the circuit is when you alter the two resistance values. The final circuit shows that you can use the thermistor properties to illuminate an LED as the temperature increases.

This circuit module could be used in many different applications when you want to be able to sense a change in temperature and activate an output circuit. For example, with some additional circuitry added to the output, it could be connected to a fan, which would be used to cool down the area when it gets too hot. This situation occurs inside a computer—you can hear the internal fan switch on when the ICs are running at a high temperature.

CHAPTER 10

Beep! Beep! Make an Electronic Horn

SUPPOSE YOU'RE DRIVING DOWN THE ROAD IN your car and you need to warn another car or a pedestrian that you're approaching. You wouldn't lean out of the window and shout, "Get out of the way!" Instead, every car comes complete with an electronic horn, which is normally situated behind the front grill of the car. The horn is activated by a switch that is usually positioned in the center of the steering wheel. Pressing the steering wheel will activate the horn, which normally makes a noise that's loud enough for other cars and people to hear.

This experiment explores how you can build a miniature horn like the one shown in Figure 10-1.

FIGURE 10-1 The electronic horn experiment

You could use this experiment to create a horn for your bike.

Experiment 7
Making an Electronic Horn

You can make electronic sounds and noises in many ways, and a few of these are discussed in this book. If you completed the experiment in Chapter 7, you made an electronic telephone and used a crystal earpiece that let you to capture and reproduce the sound of a friend's voice. A crystal earpiece will not create a noise loud enough for a horn, so you'll need to use a different component for this experiment. Figure 10-2 shows three different electronic components that you can use to create some serious noise. These are an electronic buzzer, a piezoelectric disc, and a small loudspeaker.

All the electronics required to create noise are normally inside the enclosure of an electronic buzzer. You can simply connect the buzzer to a suitable battery source and it will make a loud buzzing noise. The two other devices in Figure 10-2 require additional electronic circuitry to create noise: the piezoelectric disc, which you won't be using in this book, and the small loudspeaker that you will be using in this experiment.

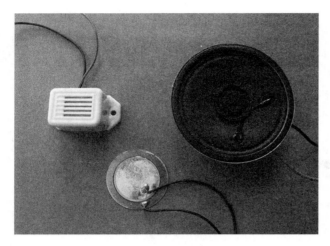

FIGURE 10-2 Various noise-making components; from left to right: electronic buzzer, piezoelectric disc, and a small loudspeaker

You can buy premade electronic horns that make a really loud sound, and this is the type of horn that car manufacturers tend to use. But that's no fun—why buy a horn when we can make one from scratch using just six components?

What Is a Loudspeaker?

This experiment uses a loudspeaker like the one shown in Figure 10-2. If you have been following the experiments in order, this will be the first time that you will have used a speaker. Loudspeakers are used to broadcast sounds in radios, sound systems, and TVs. The circuit symbol for a loudspeaker is shown here:

A speaker normally has two connections, which are sometimes marked + and − to show which side you need to connect to the positive and negative points in a circuit, respectively. Inside the speaker is a coil of wire that acts as an electromagnet (similar to, but not the same as, a relay, which you learned about in Chapter 8). When a signal is presented to the speaker connections, the changing signal activates the coil and this in turn creates vibrations in the speaker, which in turn vibrates the air and thus creates a noise that you can hear.

The Circuit Diagram

The circuit diagram for the electronic horn experiment is shown in Figure 10-3. Look at it carefully and compare it with the circuit diagram for Experiment 2 in Chapter 5. It should look familiar to you.

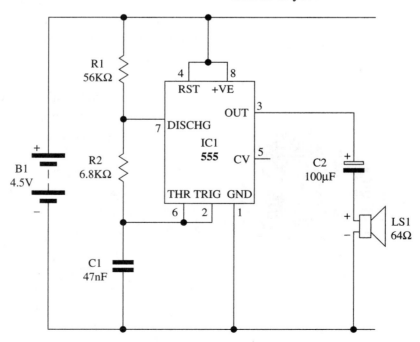

FIGURE 10-3 The circuit diagram for the electronic horn

How the Circuit Works

This circuit is made up of three building blocks: power supply, control circuitry (555 astable pulse generator), and output. The circuit is powered by three 1.5 volt AA batteries that are wired in series to create a 4.5 volt power supply. The control circuitry should be familiar to you if you have read Chapter 5, where you learned how to make a flashing LED. The control circuit contains a 555 timer (IC1) configured in astable mode, and its on-off time is set by the values of R1, R2, and C1. The capacitor (C2) and the speaker (LS1) create the output part of this circuit.

The heart of the circuit is a 555 timer in astable mode. You know that the timing of the 555 timer's output depends on the component values of the two resistors (R1 and R2) and the capacitor (C1). The timing of this circuit produces a very fast train of pulses that lets us create a noise. In previous experiments that used this method, the timing pulses have been fairly slow, at around 1Hz; in this experiment, the timing sequence pulses at around 860Hz (860 times per second). The output pin 3 feeds these pulses into a capacitor (C2), which then feeds into the speaker. The purpose of the capacitor (C2) is to remove the direct current (DC) element of the signal and provide an output that the speaker can convert into noise. This fast pulsing output vibrates the speaker at such a rate to create a loud buzzing noise that emulates a car horn.

Things You'll Need

The components and equipment that you will need for this experiment are outlined in the following table. Prepare the items that you need before starting the experiment.

Code	Quantity	Description	Appendix Code
IC1	1	555 timer	18
R1	1	56KΩ 0.5W ±5% tolerance carbon film resistor	—
R2	1	6.8KΩ 0.5W ±5% tolerance carbon film resistor	—
C1	1	47nF ceramic disk capacitor (minimum 16 volt rated)	—
C2	1	100μF electrolytic capacitor (minimum 16 volt rated)	—
LS1	1	Small loudspeaker (64Ω 0.3 watt rated)	22
B1	1	4.5V battery holder	15
B1	3	1.5V AA batteries	—
B1	1	9V PP3 battery	—
B1	1	PP3 battery clip	17
—	1	Breadboard	2
—	—	Wire links	—

NOTE

The Appendix Code column of the table refers to specific parts that I used in this experiment. Information about sourcing these parts is outlined in the Appendix.

The Breadboard Layout

Using the components in the table, build the circuit by following the breadboard layout for the electronic horn shown in Figure 10-4.

NOTE

Refer to Chapter 3 for building breadboard layouts and fault-finding guidelines.

Notice that capacitor C2 is an electrolytic device, so it needs to be connected in the circuit the correct way around.

Your speaker may not have cables connected to it; if this is the case, you can simply use some insulated solid copper wire—the type that you use for wire links—and twist the cable so that it wraps around the two speaker connections, as shown in Figure 10-5. You can then twist the two wires together, which helps to secure the two cables in place.

Figure 10-6 shows a close-up image of the breadboard layout taken at a different angle, which will help you to build the circuit.

FIGURE 10-5 Wrap the wire around the speaker connections like this.

Time to Experiment!

After you have built the circuit, you can connect the 4.5 volt battery to the circuit, as shown in Figure 10-7. The speaker should then make a loud buzzing noise.

If this works, you could try to power the circuit from a 9 volt PP3 battery instead. The noise becomes a lot louder using a 9 volt battery—not as loud as a real car horn, but still fairly loud.

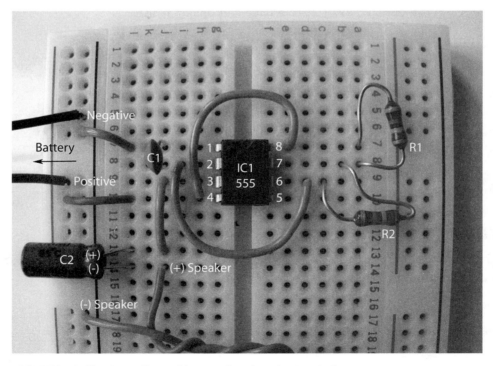

FIGURE 10-4 The breadboard layout for the electronic horn

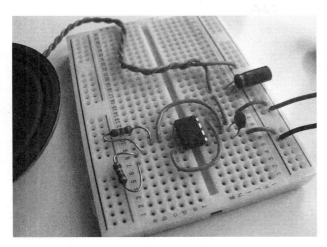

FIGURE 10-6 A close-up of the experiment around the 555 timer

Further Experimentation

After you have finished annoying your parents with the horn, you can experiment further by changing the values of R1, R2, and C1 to see what effect this has on the type of noise that is generated. You can use Table 10-1 to record your results. The formula for calculating the astable time period and frequency of the output pulses is outlined in Chapter 5.

 **HINT!**

Try to use resistor values that are higher than 1KΩ.

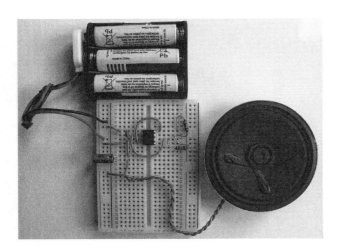

FIGURE 10-7 Connect the battery to the circuit.

Summary

In this chapter, you learned that the output from a pulsed 555 astable circuit can be used to drive a loudspeaker and create a noise. If you experimented further, you saw that by changing the frequency of the 555 timer, you can affect the frequency of the noise that is created by the speaker. You will be exploring this concept in more detail in Chapter 20.

TABLE 10-1 Record the Results of Altering the 555 Timer's Components

R1	R2	C1	On (secs)	Off (secs)	Total Time (secs)	Frequency (Hz)	Type of Noise Created
56KΩ	6.8KΩ	47nF					Buzzing noise, sounds similar to a car horn.

CHAPTER 11

Is It Raining Outside? Set Up a Water Sensor

IF YOU ARE IN A CAR IN THE RAIN, you can press a button or move a switch and the windshield wipers will turn on and wipe the rain off the windshield so that you can see where you are going. Some cars come complete with special electronic sensors that sense when it is raining and switch the wipers on automatically. This experiment shows you how to build a simple circuit and a water sensor that will illuminate an LED when water touches the probes.

Experiment 8
Setting Up a Water Sensor

This experiment shows you how a transistor is able to magnify a small current flow into a much larger one. Water will conduct electricity; however, it is not a great conductor of electricity unless the voltage or current is high enough (but not too high). You will discover that if you reduce the distance that electricity has to travel through water, you can use this to your advantage when creating a low-voltage sensing circuit like this one.

BE CAREFUL!

Never try to use water to create an electronic circuit with high voltages. This can be very dangerous.

The Circuit Diagram

The circuit diagram for the water sensor experiment is shown in Figure 11-1.

How the Circuit Works

The water sensor circuit is made up of four building blocks: power supply, input, control circuitry (NPN transistor switch), and output.

This circuit is powered by three 1.5 volt AA batteries that are wired in series to create a 4.5 volt

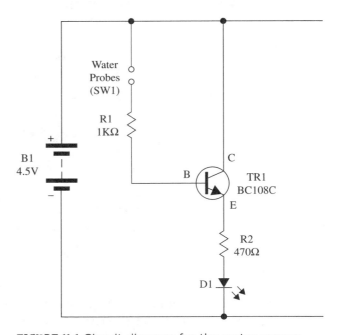

FIGURE 11-1 Circuit diagram for the water sensor

power supply. The water probes (SW1) create the input for this circuit. The NPN transistor switching circuitry is based around components R1 and TR1. The LED (D1) and its series resistor (R2) create the output part of this circuit.

The circuit is very simple and contains only four components: two resistors, a transistor, and an LED. Imagine for a moment that the two points marked "Water Probes" are actually linked together so that the positive battery rail is allowed to flow through resistor R1. If you connected the battery to the circuit, a small current would flow through resistor R1 and activate the base connection of the NPN transistor TR1. When this happens, the transistor switches on and current flows through the base and out through the emitter. This also has the effect of allowing a much larger current to flow through the collector/emitter junction of TR1, which is enough to switch on the LED.

Making the Water Probes

Before you start to build the breadboard layout for this experiment, you need to make a set of water probes. These are easy to make; the probes that I made are shown in Figure 11-2. The water probes are simply two lengths of solid insulated copper wire that have been stripped and joined together using some tape.

To make your water probes, follow these steps:

1. Cut two lengths of 2-foot-long copper wire.

2. Using your wire strippers, cut about 1 inch of insulation off the end of each length of wire— but don't remove the insulation from the cable fully. Instead, remove most of the insulation, but leave around 0.25 inch of insulation at the end of each cable.

3. Place the two pieces of wire parallel to each other and, using some insulation tape or clear

Things You'll Need

The components and equipment that you will need for this experiment are outlined in the following table. Prepare the items that you need before starting the experiment.

Code	Quantity	Description	Appendix Code
TR1	1	BC108C NPN transistor (or any high-gain transistor such as a BC267B)	9
R1	1	1KΩ 0.5W ±5% tolerance carbon film resistor	—
R2	1	470Ω 0.5W ±5% tolerance carbon film resistor	—
D1	1	5mm green LED	5
B1	1	4.5V battery holder	15
B1	3	1.5V AA batteries	—
B1	1	PP3 battery clip	17
—	1	Breadboard	1
—	—	Wire links	—
—	2	Longer wire links	—
—	—	Clear adhesive tape or electrician's insulation tape	—

NOTE

The Appendix Code column of the table refers to specific parts that I used in this experiment. Information about sourcing these parts is outlined in the Appendix.

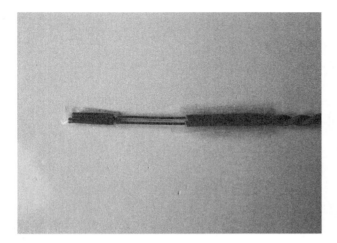

FIGURE 11-2 Homemade water probes

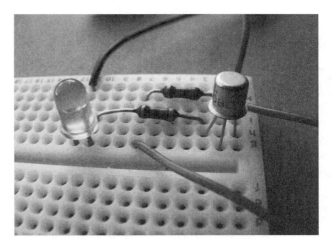

FIGURE 11-4 The breadboard layout from a different angle

tape, strap the two cables together. You should now end up with a water probe that looks like Figure 11-2, with two pieces of bare wire separated by a 1mm gap. Make sure that the bare pieces of wire are not touching each other.

4. Twist the rest of the cables together by hand to help secure the cables.

5. Remove a small piece of insulation from the other end of each piece of cable so that they can be pushed into the holes of the breadboard.

The Breadboard Layout

The breadboard layout for this experiment is shown in Figures 11-3 and 11-4, which is taken at a slightly different angle so that you can see where

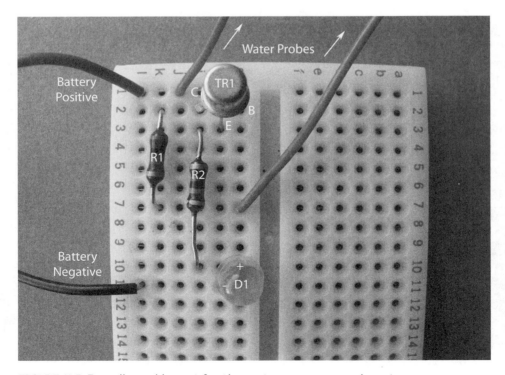

FIGURE 11-3 Breadboard layout for the water sensor experiment

the component pins are positioned. It's very simple and should take only a few minutes to build.

NOTE

Refer to Chapter 3 for building breadboard layouts and fault-finding guidelines.

Time to Experiment!

Once you have checked your breadboard layout, you can connect the battery to the circuit. You should find that the LED (D1) is not illuminated at this stage. Now moisten your finger and thumb slightly and gently touch the water probes, as shown in Figure 11-5. The LED should illuminate.

When you place your moist fingers across the water probes, a tiny current of around 10 micro amps (μA) is allowed to flow from the positive side of the battery, across your skin, and through to the base of the transistor. This current is then amplified through the transistor and allows a larger current of around 4 milliamps (mA) to flow through the collector/emitter junction of the transistor, and this illuminates the LED.

If the LED lights, you can move your experiment to the kitchen sink and open the faucet slightly so that a small stream of water runs into the sink.

FIGURE 11-5 Touch the water probe with your moist fingers.

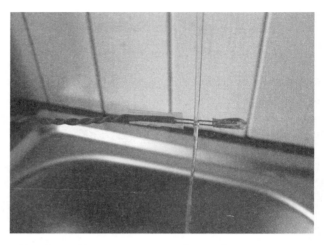

FIGURE 11-6 Water runs over the probes

Position your homemade water probes underneath the faucet, *making sure that the breadboard and battery are not going to be splashed by the water*. Now let the water run over the probes, as shown in Figure 11-6.

The LED will illuminate every time water bridges the gap between the two pieces of wire on the probes. If you move the probe in and out of the stream, the LED will flash on and off.

Further Experimentation

If you feel really adventurous, you might want to try to use this basic circuit to create an experimental water level sensor that illuminates a row of four or more LEDs to show how full a container of water is. You could do this by duplicating the circuit four times on a piece of breadboard, making sure that the four LEDs are positioned in a row. You would then need to make four water probes, which you can attach to a thin piece of wood at various positions along its length, making sure that the eight water probe cables are long enough to keep the breadboard away from the water. If you then put this water probe into a bowl of water you should find that only the water probes that touch the water will activate their associated LEDs. The higher the water level, the more LEDs will be illuminated.

Summary

You have learned how to make a switch using the amplifying properties of a bipolar transistor. This, in conjunction with the fact that a small current is able to flow across a drop of water, demonstrates how a few components can be used to sense the presence of water and switch on an LED. This shows you how a rain sensor in a car might work to operate the windshield wipers rather than illuminating an LED. A similar unit in a car would need additional circuitry to activate the wiper motor and possibly some timing circuitry.

You will be building on the concepts that you have learned in this chapter in an experiment in Chapter 14, when you put this circuit module to good use and build a touch-activated alarm.

PART FOUR

Spy vs. Spy!
Security Experiments

CHAPTER 12

Intruder Alert! Design a Basic Alarm Circuit

THIS PART OF THE BOOK CONTAINS experiments that explore security circuits. There are many types of security products on the market today, such as home and vehicle burglar alarms, home security lights that switch on when it is dark, and security keypads that require you to press a sequence of buttons to gain entry into a security door, to name a few. You'll be exploring all these security products in this section of the book and building experimental circuits that emulate some of their features. You can use Experiment 9 as the basis of your own alarm project to protect your belongings in your bedroom and warn you if grown-ups or your brother or sister enter your room.

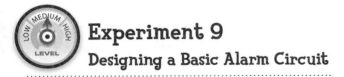

Experiment 9
Designing a Basic Alarm Circuit

In this experiment, you will build a basic alarm circuit that will illuminate an LED when a set of contacts is made. In Chapter 13, you'll learn how to modify this experiment to generate an audible alarm rather than an LED output.

Here's a typical specification for a basic burglar alarm:

- **Normally open contacts**   Contacts could be connected to a pressure-sensitive mat that activates when someone stands on it.
- **Output connections**   These can be connected to an audible output device.
- **Latching alarm**   This type of alarm stays on once triggered, even though the person may have stepped off the pressure-sensitive mat.
- **Time delay**   Once an alarm is triggered, it will eventually switch off the sound after a predetermined time; this will be long enough to frighten the burglar away but not so long that it become a nuisance to other people.

You could build an electronic circuit in a number of different ways to create this specification. In this experiment, you are going to use the versatile 555 timer integrated circuit (IC) once again, but this time you will be using its second operating mode to create your circuit: *monostable* mode.

555 Timer: Monostable Mode

If you have followed each chapter in sequence and built each experiment, you will have already built the LED flasher in Chapter 5, which used the 555 timer in astable mode. The beauty of the 555 timer chip is that it also has a second mode of operation: *monostable* mode. You have already learned that an astable circuit produces a pulsed output signal that you can use to flash an LED on and off. A monostable circuit produces a "one-shot" output signal that lets you illuminate an LED for a length of time, and this is ideal for this experiment. The circuit diagram for a basic 555 monostable timer circuit is shown in Figure 12-1.

The circuit diagram shows the 555 timer wired in a configuration slightly different from the astable circuit in Chapter 5. This time, only one resistor R1 and one capacitor R2 are used to create the timing sequence, and the wiring of the circuit is slightly different from that of the astable circuit. A different formula is used for calculating the length of time that the output remains on when the trigger (pin 2) is connected to the negative battery connection:

Output Time (in seconds) = $1.1 \times R1 \times C1$

So, for example, if you used a resistor (R1) with a value of 22KΩ and a capacitor (C2) with a value of 100μF, you can use these values in the monostable formula:

$$Output\ Time = 1.1 \times 22000Ω \times 0.0001F$$
$$= 2.42\ seconds$$

This means that an LED connected to the output pin would be illuminated for about 2.5 seconds after the timer has been triggered. The accuracy of this time period is determined by the accuracy of the components that you use, so in reality the time period may vary slightly from your calculated value.

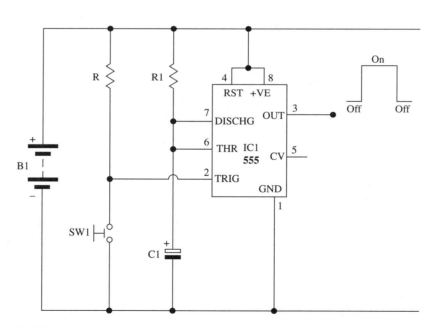

FIGURE 12-1 A basic 555 monostable timer circuit

How to Convert Capacitor and Resistor Values into Numbers That You Can Use in the Formulas

You need to convert the capacitor value to farads and the resistor value to ohms when using the monostable formula. You can use the following calculations to help you do this.

Capacitance

- To convert picofarads (pF) to farads, *divide* the pF value by 1,000,000,000,000.

- To convert nanofarads (nF) to farads, *divide* the nF value by 1,000,000,000.

- To convert microfarads (μF) to farads, *divide* the μF value by 1,000,000.

Resistance

- To convert kilo-ohms (KΩ) to ohms (Ω), *multiply* the KΩ value by 1000.

- To convert mega-ohms (KΩ) to ohms (Ω), *multiply* the MΩ value by 1,000,000.

The Circuit Diagram

Let's take a look at the circuit diagram for the basic alarm circuit, shown in Figure 12-2.

How the Circuit Works

The circuit diagram should look familiar, because it is based on the 555 monostable timer circuit shown in Figure 12-1.

The basic alarm circuit is made up of four building blocks: power supply, input, control circuitry (555 monostable), and output. This circuit is powered by three 1.5 volt AA batteries that are wired in series to create a 4.5 volt power supply. The normally open alarm input switch (SW1) and the resistor (R2) create the input for this circuit. The 555 monostable circuit based around the 555 timer IC (IC1), resistor R1, and capacitor C1 is the control part of the circuit. You will also notice that an additional capacitor (C2) has been connected between pin 5 and the negative battery connection; this is to stop output pin 3 from triggering unexpectedly, which could occur without it. The LED (D1) and its series resistor (R3) create the output part of this circuit.

Here's how the circuit works: When the battery is connected to the circuit, the resistor R2 provides a positive voltage to the trigger input of IC1 (pin 2), which means that the output of the 555 timer (pin 3) is switched off. In this state, the alarm circuit is

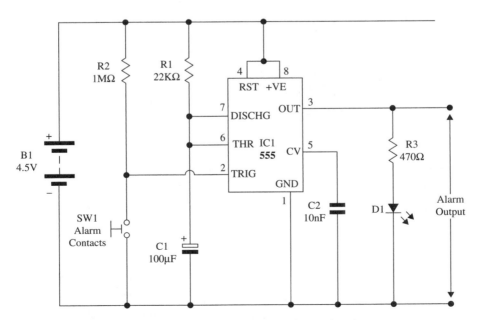

FIGURE 12-2 The circuit diagram for the basic alarm circuit

waiting for the alarm input (SW1) to be activated. If this happens, the trigger pin (pin 2) is connected to the negative battery connection, which causes the monostable timing sequence of the 555 timer to start. This causes the output to switch on for a period of time, which is determined by the values of the resistor (R1) and the capacitor (C1). The output is connected to an LED (D1) and its series resistor (R3).

The Breadboard Layout

The breadboard layout for this experiment is shown in Figure 12-3. It also includes the component codes, which help you to identify each of the components.

Figure 12-4 shows the breadboard layout from a different angle so that you can see more of the pin configurations.

Things You'll Need

The components and equipment that you will need for this experiment are outlined in the following table. Prepare the items that you need before starting the experiment.

Code	Quantity	Description	Appendix Code
IC1	1	555 timer IC	18
R1	1	22KΩ 0.5W ±5% tolerance carbon film resistor	—
R2	1	1MΩ 0.5W ±5% tolerance carbon film resistor	—
R3	1	470Ω 0.5W ±5% tolerance carbon film resistor	—
C1	1	100µF electrolytic capacitor (minimum 10 volt rated)	—
C2	1	10nF ceramic disk capacitor (minimum 10 volt rated)	—
D1	1	5mm green LED	5
SW1	2	Wire links	—
B1	1	4.5V battery holder	15
B1	3	1.5V AA batteries	—
B1	1	PP3 battery clip	17
—	1	Breadboard	2
—	—	Wire links	—
—	—	Various 0.5W ±5% tolerance carbon film resistors and capacitors (minimum 10 volt rated)	—

NOTE

The Appendix Code column of the table refers to specific parts that I used in this experiment. Information about sourcing these parts is outlined in the Appendix.

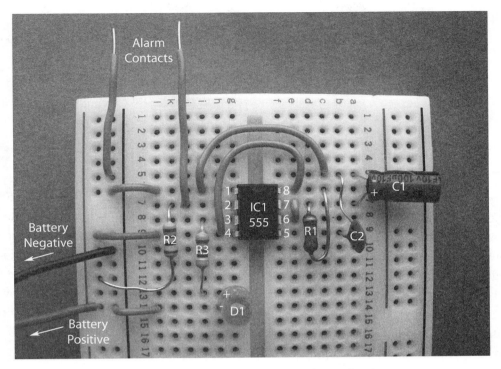

FIGURE 12-3 The breadboard layout for the basic alarm circuit

Time to Experiment!

Once you have built the breadboard layout, connect the battery to the circuit. You should find that the LED does not illuminate. Now touch the two ends of the "alarm contact" wires together for a second, as shown in Figure 12-5, and the LED will illuminate for a short time.

If you use a watch to time how long the LED illuminates, you should find that it stays lit for around 2.5 seconds, as per the calculation earlier in this chapter. After this time, the LED switches off again and the circuit waits for the alarm contact wires to be connected together again. If you keep the two alarm contact wires connected together, the LED will remain illuminated until the battery runs dry.

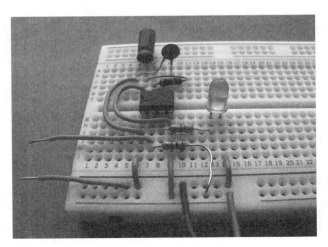

FIGURE 12-4 The breadboard layout taken from a different angle

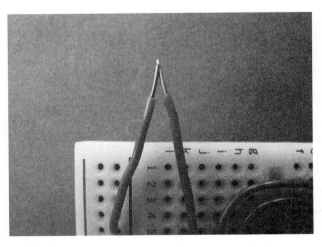

FIGURE 12-5 Touch the two alarm contact wires together briefly.

TABLE 12-1 Monostable Timing Table

R1	C1	LED On Time
22KΩ	100μF	2.42 seconds

Try to experiment with the values of the resistor R1 and the capacitor C1 to see how this affects the length of time that the LED remains illuminated after the trigger pin has been activated. You can use the spaces in Table 12-1 to write down the component values that you try out and the length of time that the LED is illuminated. The top row shows the component values used in this experiment. You could also use the monostable timing formula shown earlier; don't forget that you need to convert the value of the resistor into ohms and the value of the capacitor into farads. When you have finished experimenting with the component values, try to find a resistor/capacitor combination that creates a time period of approximately 25 seconds, because this will help you in the next experiment.

HINT!

Experiment with resistors that have values between 1KΩ and 100KΩ, and capacitors that have values between 1μF and 1000μF.

Summary

In this chapter, you learned how the 555 timer chip can be configured in its monostable mode to create a predetermined timed output. You also learned that you can alter the amount of time that the LED is illuminated by altering two component values. This type of circuit is very versatile, and you could use it at the heart of many different timing circuits—for example, a timer for boiling an egg.

The basic alarm circuit that you have built is fine as a basic alarm, but illuminating an LED is not going to be enough to deter a thief. You will need to find a suitable method of triggering the circuit, rather than just touching two pieces of wires together. In the next chapter, you will discover how to modify the basic alarm circuit to incorporate additional features.

CHAPTER 13

Step Inside! Make a Pressure-Sensitive Mat

THIS CHAPTER SHOWS YOU HOW YOU CAN modify the basic alarm circuit that you built in the last experiment to make it work more like a proper burglar alarm. Before you start this experiment, you need to have read Chapter 12 and built the experimental alarm circuit, because you will be using it as the basis of this experiment.

Experiment 10
Making a Pressure-Sensitive Mat

In Experiment 9, you built a basic alarm circuit that illuminated an LED for a 25-second period when a pair of alarm contacts was connected. The circuit used a 555 timer in monostable mode. The circuit diagram for the basic alarm circuit is shown in Figure 13-1 and the breadboard layout is shown in Figure 13-2.

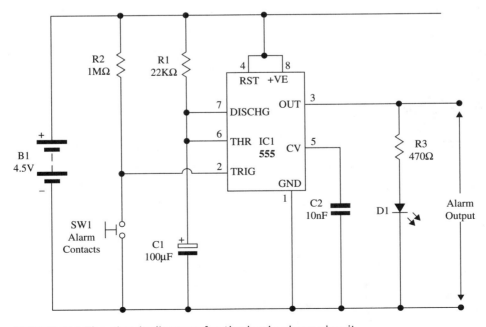

FIGURE 13-1 The circuit diagram for the basic alarm circuit

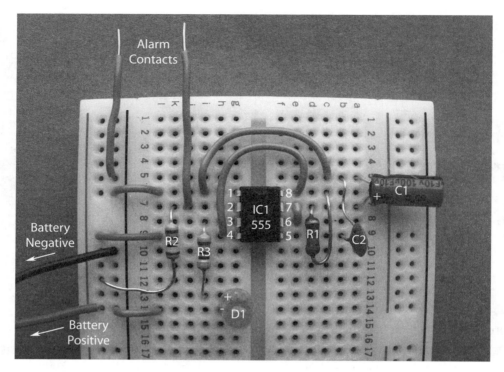

FIGURE 13-2 The breadboard layout for the basic alarm circuit

Additional Features

To make the circuit more like a real burglar alarm you need to add two features to the experiment:

- Make a pressure-sensitive mat that can be connected to the alarm contacts. This could be positioned inside your bedroom door, so that when someone steps on it, the alarm will be triggered. (You could buy a premade pressure-sensitive mat from a shop that sells security products, but where's the fun in buying one when you can build it instead?)

- Replace the LED with a component that makes a noise when the alarm is activated. A loud noise will be more of a deterrent to a would-be burglar than just an LED.

These two additional features will be explored in this chapter.

Things You'll Need

The household objects and components that you will need for this experiment are outlined in the following table. Prepare the items that you need before starting the experiment.

Code	Quantity	Description	Appendix Code
—	1	A normal mat or rug	—
—	1	A piece of thick cardboard	—
—	1	Aluminum foil (the typical household type)	—
—	1	Cling film (the clear plastic type used in the kitchen)	—
—	1	Double-sided adhesive sponge strips	26
—	1	Single-sided clear adhesive tape	—
—	2	2-foot length of wire	—

Things You'll Need (continued)

Code	Quantity	Description	Appendix Code
—	1	6 volt buzzer	25
D2/D3	2	1N4003 diode	—

NOTE

The Appendix Code column of the table refers to specific parts that I used in this experiment. Information about sourcing these parts is outlined in the Appendix.

Build the Mat

The following steps and illustrations explain how to build the pressure-sensitive mat.

1. Cut two pieces of cardboard approximately 3-by-3 in. square (75-by-75mm).

2. Wrap each piece of card tightly with a sheet of aluminum foil.

3. Strip the insulation from two long pieces of wire, and use clear tape to attach the bare wire to the rough side of each piece of aluminum-covered cardboard.

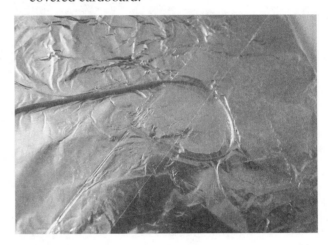

4. Cut five lengths of adhesive sponge strips and apply them to the smooth side of one piece of aluminum-covered cardboard. Then attach a second layer of strips on top of the first layer to double the thickness.

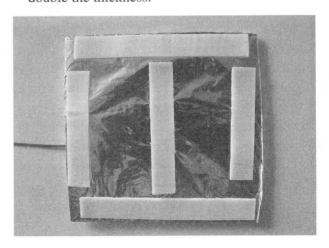

5. Press the other piece of aluminum-covered cardboard onto the adhesive strips so that the two pieces of cardboard stick together. The two pieces of aluminum should now be separated from each other by two layers of adhesive strips.

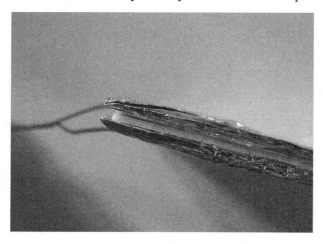

6. Twist the two wires together to make the connection stronger, and wrap the complete assembly with some cling film so that the pad is totally insulated.

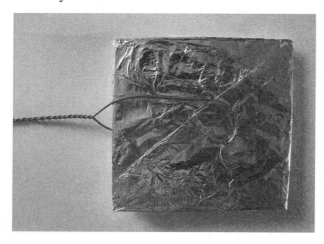

You have now created your pressure-sensitive pad!

You can test your pressure-sensitive pad by placing it underneath a rug or a mat and connecting the two wires to your multimeter on the continuity setting. If you step on the mat, the multimeter should bleep, showing that electrical contact has been made (Figure 13-3). This happens because your weight pushes the two pieces of aluminum foil together so that they touch each other.

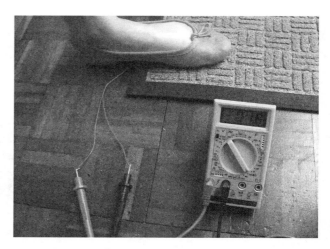

FIGURE 13-3 Test that the pressure pad works using your multimeter.

If you step off the mat, the meter should stop bleeping, because the sponge pads push the two pieces of aluminum foil apart. Try walking on and off the mat to make sure that it works as expected.

 BE CAREFUL!

The pressure mat that you have made is able to switch only very small currents like those used in this experiment. Do not try to use the pressure pad to operate buzzers or LEDs directly.

The Circuit Diagram

Now that you have built the mat, the next stage is to modify the original alarm circuit slightly so that you can replace the LED with a buzzer. The modified circuit diagram is shown in Figure 13-4.

How the Circuit Works

Notice that the modified circuit diagram no longer contains the LED and its series resistor; these components have been replaced with two diodes that are configured so that they can operate a 6 volt buzzer. If you compare this output configuration to the circuit diagram for the indicator lights in Chapter 8, you will notice that this diagram is

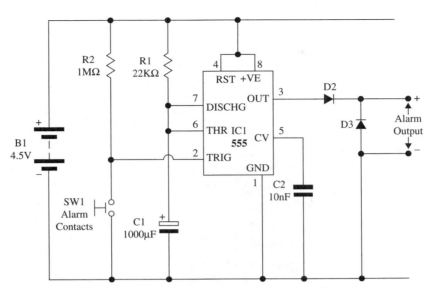

FIGURE 13-4 The circuit diagram for the modified alarm circuit. Notice how the value of C1 is 1000µF in this circuit which creates a 25-second time period.

similar to the circuit that operates the relay coil. The 555 timer is able to switch enough current to drive a small buzzer directly, and the two diodes are required to operate it correctly. The rest of the circuit operates in the same way as the previous experiment. Whenever the pressure-sensitive mat is activated, it triggers the alarm contacts and operates the 555 monostable circuit, and the output of this circuit operates the buzzer for 25 seconds.

 NOTE

The 555 timer IC used in this experiment is able to activate a buzzer directly so long as the current rating of the buzzer is less than 200mA; some variants of the 555 timer can only switch up to 100mA; therefore, you need to make sure that your buzzer has a current rating lower than 100mA. Experienced electronics designers and inventors might want to design electronic circuits that switch higher currents; to do this they might decide to use a relay to switch the device instead (you can read more about relays in Chapter 8).

The Breadboard Layout

The modified breadboard layout is shown in Figure 13-5. Notice that the LED and resistor have been replaced with the two diodes, D2 and D3. The 6 volt buzzer can then be connected to the breadboard as shown.

The two cables of the pressure-sensitive mat can be plugged in place of the alarm contact wires. The final experiment should look like Figure 13-6.

Time to Experiment!

Now that the circuit is complete, you can try it out. Connect the batteries to the circuit, and you should find that nothing happens yet. Either step on the pressure pad or press on it using your fingers; this should activate the buzzer for a period of 25 seconds.

If this works as expected, you can set up the burglar alarm in a place that you want to protect. For example, you can hide the pressure pad under a mat in your bedroom. Switch on the alarm when you leave your room, and then, if someone enters your room and steps on the mat, he or she will get a big surprise. The buzzer will make a loud noise

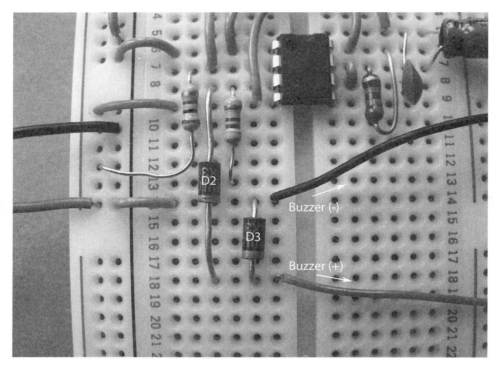

FIGURE 13-5 The modified breadboard layout

that will frighten away the intruder, and it will also warn you that someone has entered your room so that you can investigate. It won't matter that they have stepped off the mat, because the monostable circuit that you built will continue to operate the buzzer for 25 seconds!

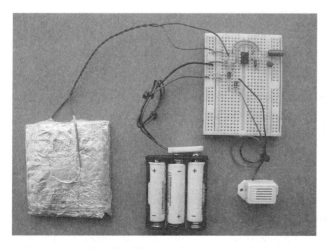

FIGURE 13-6 The final layout for the burglar alarm

 HINT!

You can connect the pressure mat to some longer cables so that the breadboard is positioned further away from the mat.

Summary

In this experiment, you learned how to build a pressure-sensitive burglar alarm using some common household objects: you created a normally open momentary switch. Because of the simple materials and methods you used to make the switch, you may not realize that you have also created a capacitor.

If you read Chapter 3, you know that a capacitor is basically two metal plates that are separated by an insulating material. Think about the pressure pad: you used two aluminum plates separated by a sponge tape, which creates an air gap and this acts as an insulator. Figure 13-7 shows a pressure pad attached to the capacitor connections of my multimeter. The reading shows that the pressure pad has a capacitance of around 45pF, which is a very low capacitance.

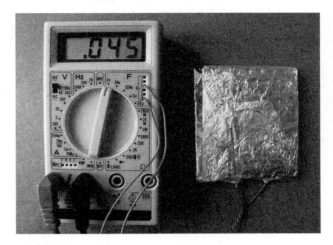

FIGURE 13-7 Your pressure pad also has a capacitance rating!

CHAPTER 14

Careful What You Reach For! Create a Touch-Activated Alarm

IN EXPERIMENT 10, YOU LEARNED HOW to make a pressure-sensitive mat and connect it to an alarm circuit that triggers when someone stands on the mat. But what happens if the crafty burglar or snooper manages to get past your pressure-sensitive mat without setting the alarm off, and is then heading for your private stuff, such as your journal? You'll need to find another method of security to stop the snooper in his or her tracks.

This experiment shows you how to make an alarm circuit that's triggered by the crafty snooper touching an object. The touch-activated alarm mounted in a special safe box is shown in Figure 14-1.

FIGURE 14-1 The touch-activated safe box

Experiment 11
Creating a Touch-Activated Alarm

This experimental circuit shows you how you put together different circuit modules to create a slightly more complex circuit design. As you learn more about how electronic circuits work, you will find it easy to join circuits together. Soon you'll become an expert inventor!

The Circuit Diagram

The circuit diagram for the touch-activated alarm is shown in Figure 14-2.

How the Circuit Works

Compare the circuit diagram shown in Figure 14-2 with the alarm circuit for the pressure-sensitive mat in Chapter 13. It looks very similar. Now look at the circuit diagram for the water sensor experiment in Chapter 11, and you will see that this circuit is also built in to the circuit diagram for this experiment.

The touch-activated alarm circuit is made up of four building blocks: power supply, input, control circuitry (555 monostable), and output. This circuit is powered by three 1.5 volt AA batteries that are

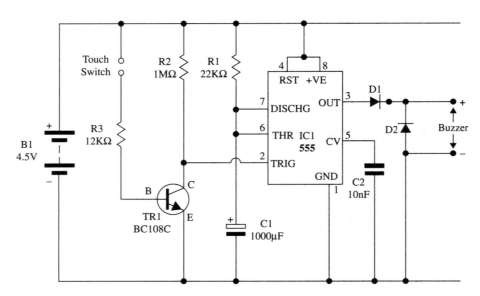

FIGURE 14-2 The circuit diagram for the touch-activated alarm

wired in series to create a 4.5 volt power supply. The touch switch contacts, transistor TR1, and resistors R2 and R3 create the input part of this circuit. The transistor acts as an amplifier in this circuit. The 555 monostable circuit based around the 555 timer IC (IC1), resistor R1, and capacitor C1 is the control part of the circuit. Capacitor C2 has been connected between pin 5 and the negative battery connection; this will stop output pin 3 from triggering unexpectedly, which could occur without it. The two diodes, D1 and D2, and the 6 volt buzzer create the output part of this circuit.

You should be getting familiar with the way this circuit works. The alarm part of the circuit is a 555 monostable timer circuit that activates the 6 volt buzzer for around 25 seconds when it is triggered. This timing period is set by the values of resistor R1 and capacitor C1—you read about monostable timing periods in Chapter 12.

To activate the 555 monostable timer, the trigger pin (pin 2) needs to be connected to the negative side of the battery. In this circuit resistor, R2 ensures that pin 2 is connected to the positive rail, which means that the 555 timer does not trigger until it's supposed to.

Remember that we are trying to make a monostable circuit that triggers using touch only; therefore, you need to include some additional circuitry to do this. This is where the transistor (TR1) and its base resistor (R3) are needed. Recall from Chapter 11's water sensor that this type of circuit is quite sensitive to water and causes the transistor to switch on. The circuit wasn't quite sensitive enough to illuminate an LED brightly when you touched the contacts, but it did allow a small amount of current to flow through the transistor when you touched the contacts. In the circuit we'll create here, touching the contacts allows enough current to flow through the base of the transistor to switch it on slightly, which in turn allows current to flow between the collector and emitter of TR1. This transistor switching allows the negative of the battery to reach pin 2 of the 555 timer, and because the trigger pin of the 555 timer is quite sensitive this brief pulse is enough to activate the 555 timer and sound the buzzer.

The 555 timer integrated circuit (IC) is powerful enough to activate the buzzer without your having to use another transistor amplifier to drive it, but you do need to include the two diodes (D1 and D2) in the circuit to make the buzzer work properly. D1 makes sure that the positive signal from output pin 3 reaches the buzzer. D2 is a flywheel diode that stops flyback (or back EMF) from reaching pin 3 of the 555 timer.

NOTE

You read about flywheel diodes and flyback in Chapter 8.

The Breadboard Layout

The breadboard layout photograph for this experiment is shown in Figure 14-3, which also shows each of the component codes in the parts list table.

NOTE

Refer to Chapter 3 for building breadboard layouts and fault-finding guidelines.

Things You'll Need

The components and equipment that you will need for this experiment are outlined in the following table. Prepare the items that you need before starting the experiment.

Code	Quantity	Description	Appendix Code
IC1	1	555 timer IC	18
R1	1	22KΩ 0.5W ±5% tolerance carbon film resistor	—
R2	1	1MΩ 0.5W ±5% tolerance carbon film resistor	—
R3	1	12KΩ 0.5W ±5% tolerance carbon film resistor	—
C1	1	1000µF electrolytic capacitor (minimum 10 volt rated)	—
C2	1	10nF ceramic disk capacitor (minimum 10 volt rated)	—
D1, D2	2	1N4003 diode	—
TR1	1	BC108C NPN transistor	9
SW1	2	Wire links	—
B1	1	4.5V battery holder	15
B1	3	1.5V AA batteries	—
B1	1	PP3 battery clip	17
—	1	6 volt buzzer*	25
—	1	Breadboard	2
—	1	VHS video case (or a suitable plastic box)	—
—	1	Aluminum foil	—
—	1	Double-sided adhesive sponge tape	26

*Make sure that the buzzer has a current rating less than 100mA (see the note in Chapter 13 about the current capabilities of the 555 timer IC).

NOTE

The Appendix Code column of the table refers to specific parts that I used in this experiment. Information about sourcing these parts is outlined in the Appendix.

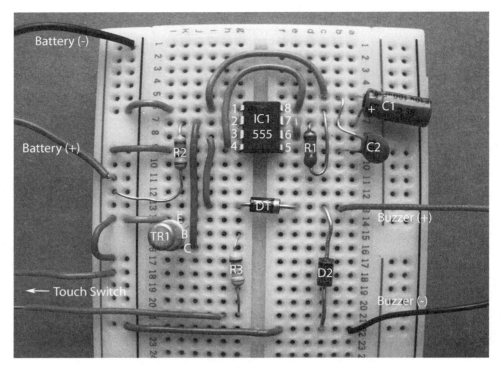

FIGURE 14-3 The breadboard layout for the touch-activated alarm

Figures 14-4 and 14-5 show the breadboard layout from different angles to provide a more detailed view; this will help you to see how the components are fitted to the breadboard.

Once the circuit has been built, you can try it out. Connect the battery to the circuit and then touch the two touch-switch wires with the end of your finger. You should find that the buzzer sounds for about 25 seconds. If this happens, you are ready to build a touch-activated safe!

Time to Experiment!

I made a "safe" using an empty VHS video case. You are probably too young to remember VHS video tapes, but your mom and dad used to watch

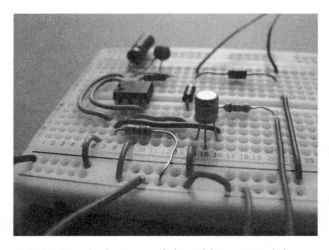

FIGURE 14-4 A close-up of the wiring around the transistor

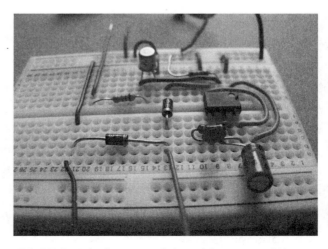

FIGURE 14-5 A close-up of the wiring around the diodes

films on video tape before DVDs were invented! The video tape's plastic cover lets you slide in a piece of paper with your own design printed on it, like the image in Figure 14-1. You might need to ask your parents to find one for you; or you can use any plastic box that's big enough for the breadboard, a battery, and your journal (or iPad, or whatever item you want to keep safe) inside.

First, you need to make two touch-sensitive plates using some aluminium foil and double-sided adhesive tape. The following steps and illustrations explain how to make the safe:

1. Strip the insulation from a 12-inch-long piece of interconnecting wire and attach it to a piece of double-sided adhesive, as shown here:

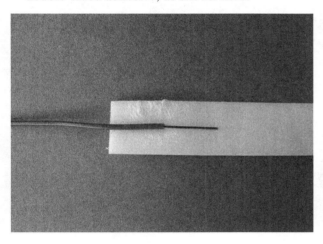

2. Attach the tape to a piece of aluminum foil so that the wire touches the foil and is sandwiched between the foil and tape, like so:

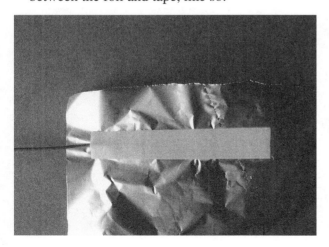

3. Trim the foil and the adhesive tape to size using scissors so that the width of the foil matches the tape. You need to make two touch-sensitive plates so repeat steps 1 and 2 to create the second plate.

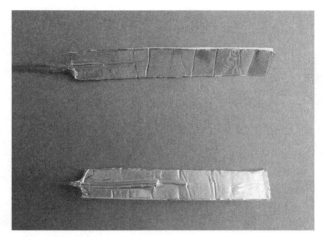

4. Attach the foil strips to the outside of the VHS box—one on either side of the box, so that if someone picks up the box, they will touch the foil strips on both sides.

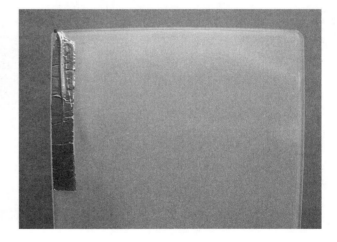

5. Fix the breadboard and battery holder inside the box using double-sided tape, making sure that there is enough room to hide your journal or belongings inside.

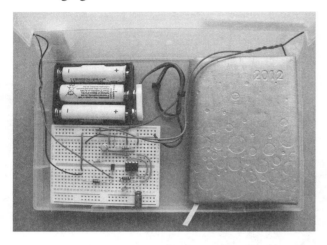

6. Fit the buzzer inside or outside the box. I chose the outside because it sounds louder.

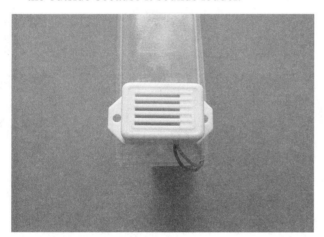

Once you have built the safe, try it out to see if it works. Connect the battery to the circuit and then close the box with the journal inside; try not to touch the aluminum foil strips. Next, see if you can open the box without setting off the buzzer. You'll find that it's very difficult to do this, because you need to touch the aluminum foil strips to open the box, and every time that you try to open the box, the buzzer makes a loud noise for 25 seconds. This is enough time to frighten away any snooper and warn you that someone is trying to find your stuff.

NOTE

The circuit is quite sensitive, and sometimes the alarm will trigger even when you touch just a single piece of aluminum foil that's connected to resistor R3.

Summary

In this experiment, you learned that electronic circuits can be constructed by attaching several electronic circuits together. This is the secret to building complex electronic circuits.

CHAPTER 15

You'll Never Get In! Build an Electronic Security Keypad

MANY SECURITY DOORS HAVE A *security keypad*, which is usually a number of push buttons that a person has to press to activate the secret code to open the door. This stops thieves and snoopers from gaining access to places they're not supposed to be. This experiment shows you one method of creating a keypad with eight push-button switches that have to be pressed in a certain way to illuminate an LED. The layout for the experiment is shown in Figure 15-1.

Experiment 12
Building an Electronic Security Keypad

This experiment demonstrates how you can wire eight simple push-button switches to create a security keypad. It also demonstrates the principle of two important *logic gates*.

INTERESTING FACT

Logic gates are electronic circuits that are the building blocks inside many digital integrated circuits (ICs). Logic gates are individual circuits that in their basic form contain two inputs and one output; the way that the output of each of these logic gates switches on or off depends on the type of logic gate used and the signals connected to the input. Logic gates inside ICs are sometimes created using a number of miniature transistors, diodes, and resistors. This family of ICs is called TTL, which stands for Transistor-Transistor Logic.

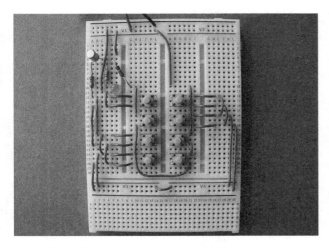

FIGURE 15-1 The electronic security keypad experiment

The Circuit Diagram

The circuit diagram for the security keypad is shown in Figure 15-2. The circuit is really simple; it comprises only five electronic components and eight push-button switches.

How the Circuit Works

The security keypad circuit is made up of four building blocks: power supply, input, control circuitry (NPN transistor switch), and output. This circuit is powered by three 1.5 volt AA batteries that are wired in series to create a 4.5 volt power supply. The eight normally open push-button switches (SW1–SW8) form the input part of this circuit. The transistor switching circuitry is based around components R1, R2, and TR1. The LED (D1) and its series resistor (R3) create the output part of this circuit.

Before the control circuitry is explained, take a closer look at the way that the eight switches are wired together. You will notice that switches 3, 6, and 8 are wired in *series*, which means that in order for the positive rail of the battery to reach

resistor R1, all of these switches have to be pressed simultaneously. This type of configuration is called an *AND* logic *circuit*, because switch 3 *AND* switch 6 *AND* switch 8 have to be pressed at the same time to operate the circuit. As you learn more about electronics you will find out that AND circuits are used in computer and digital circuits (you can read more about this in the summary).

 INTERESTING FACT

The way that a logic gate operates can be shown in the form of a table, called a truth table. The truth table shows the various ways that the input connections can be switched on or off and how this affects the output. Digital circuits only have two possible circuit states, either *on* (represented by a binary number 1) or *off* (represented by a binary number 0), and this is why digital circuits use binary numbers at their heart. Binary numbers are different from decimal numbers because they only contain numbers 0 and 1, rather than numbers 0 to 9.

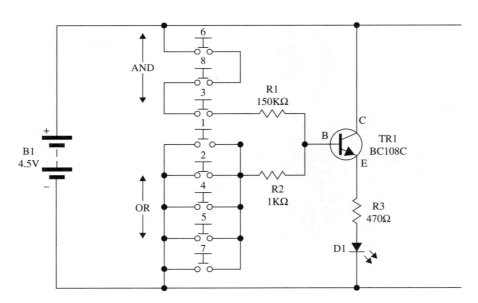

FIGURE 15-2 The circuit diagram for the electronic security keypad

TABLE 15-1 AND Logic for a Two-Switch Circuit

SW1	SW2	OUTPUT
Off (O)	Off (O)	Off (O)
Off (O)	On (1)	Off (O)
On (1)	Off (O)	Off (O)
On (1)	On (1)	On (1)

The truth table for an AND circuit with two switches is shown in Table 15-1. The truth table shows that the only time that the output switches on is when both switch 1 *AND* switch 2 is switched on; the binary number for each state is also shown in brackets.

There are two types of circuit symbol that you will come across for an AND logic gate and they look like this:

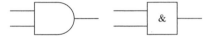

The two lines on the left of each symbol represent the two inputs, and the line on the right of each symbol is the output.

Look at switches 1, 2, 4, 5, and 7 in Figure 15-2 and you'll notice that these switches are connected up differently from the other three. These switches are wired *in parallel*, which means that in order for the negative rail of the battery to reach resistor R2, any one of the switches can be pressed. This type of configuration is called an *OR logic circuit*, because switch 1 OR switch 2 OR switch 4 OR switch 5 OR switch 7 has to be pressed to operate the circuit.

The logic table for an OR circuit with only two push-button switches is shown in Table 15-2.

TABLE 15-2 OR Logic for a Two-Switch Circuit

SW1	SW2	OUTPUT
Off (O)	Off (O)	Off (O)
Off (O)	On (1)	On (1)
On (1)	Off (O)	On (1)
On (1)	On (1)	On (1)

The truth table shows that the output switches on whenever either switch 1 *OR* switch 2 is switched on; the binary number for each state is also shown in brackets.

There are two types of circuit symbol that you will come across for an OR gate and they look like this:

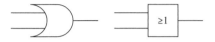

Again, the two lines on the left of each symbol represent the two inputs and the line on the right of each symbol is the output.

Now that you understand how the switch configurations work, let's look at the control circuitry.

Here's how the circuit works: Switches 3, 6, and 8 must to be pressed at the same time to allow the positive battery rail to feed through resistor R1, which drives the base of the NPN transistor and in turn illuminates the LED (D1). This is the secret code to operate the security keypad.

This in itself makes a pretty secure keypad. We'll set it up to make it even more secure, however, so that if a person presses another switch simultaneously with the other three, a circuit pathway to the negative rail of the circuit is created, and this will stop the positive signal reaching the base of the transistor, so the LED doesn't illuminate. This circuit configuration creates quite a secure keypad, as you will soon see when you build the experiment.

It's important that the resistor R1 has a high resistance, because when the secret code is pressed on the keypad, plus any other key, the battery voltage feeds through both resistors, creating a short-circuit. If the resistor has a low resistance, more current is drawn from the battery in this state, and this could cause the resistor to get too hot.

Things You'll Need

The components that you will need for this experiment are outlined in the following table. Prepare the items that you need before starting the experiment.

Code	Quantity	Description	Appendix Code
TR1	1	BC108C NPN transistor (or any high-gain transistor such as a BC267B)	9
R1	1	150KΩ 0.5W ±5% tolerance carbon film resistor	—
R2	1	1KΩ 0.5W ±5% tolerance carbon film resistor	—
R3	1	470Ω 0.5W ±5% tolerance carbon film resistor	—
D1	1	5mm green LED	5
SW1–SW8	8	6mm-by-6mm push to make a normally open switch (printed circuit board mounted type)	24
B1	1	4.5V battery holder	15
B1	3	1.5V AA batteries	—
B1	1	PP3 battery clip	17
—	1	Breadboard	3
—	—	Wire links	—

NOTE

The Appendix Code column of the table refers to specific parts that I used in this experiment. Information about sourcing these parts is outlined in the Appendix.

The Breadboard Layout

The breadboard layout for the security keypad experiment is shown in Figure 15-3.

The breadboard might look complicated, because it contains quite a few wire links. Be careful when building this experiment, and double-check to make sure your wiring matches what's shown in the figure. Refer to the close-up photographs in Figures 15-4 and 15-5, which show more details of the wiring.

Time to Experiment!

Once you have finished building the breadboard layout, connect the battery to the circuit, and you should find that the LED is not illuminated. Now press push buttons 3, 6, and 8 at the same time,

as shown in Figure 15-6. Keep them pressed, and the LED should illuminate. This is the secret code that activates the LED. In a real-life security application, this might activate a relay to unlock a door, for example.

This is the only button sequence that will activate the LED. Now try removing one or two of your fingers from any of these three buttons, and the LED no longer illuminates. This happens because these buttons are wired *in series* to form an AND gate, as explained earlier, and this means that all three buttons have to be pressed at once to complete the circuit.

Press the three buttons again to illuminate the LED, and then press any of the other buttons at the same time. The LED switches off, because these five

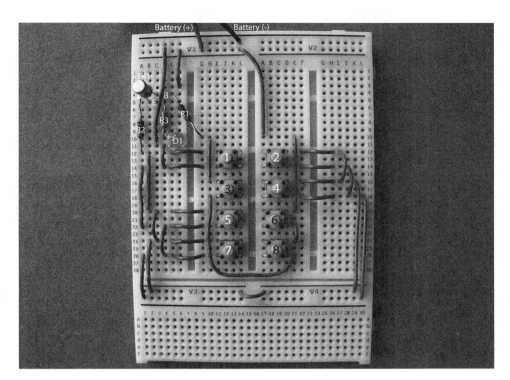

FIGURE 15-3 The breadboard layout for the security keypad experiment

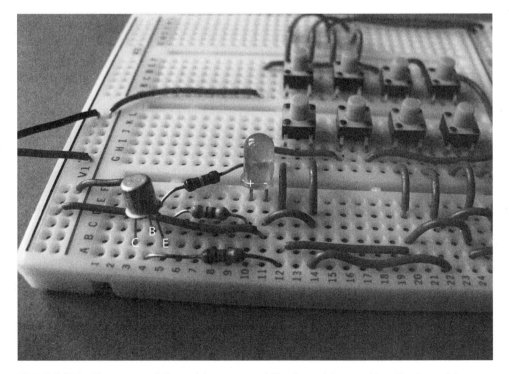

FIGURE 15-4 Close-up of the wiring around the transistor; notice the transistor and LED connections

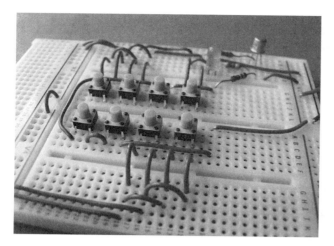

FIGURE 15-5 Close-up of the wiring around the switches

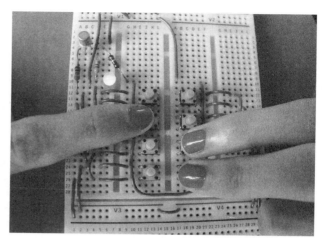

FIGURE 15-6 Pressing the secret buttons to active the LED

switches are wired *in parallel* and create an OR gate, as explained earlier in this chapter.

You have built a secret keypad that isn't that easy to crack, unless you know which three buttons to press. Even if you start to press random buttons, it will be very difficult to figure out the code and keep the LED illuminated. Try it out on your friends to see if they can crack the code. You don't need to tell them that they have to press three buttons at once; just ask them to see if they can press the buttons in a certain way to make the LED illuminate for more than 3 seconds. They may discover the secret code eventually, but it will probably take them a while to figure it out.

You might also want to play around with the switch wiring to make the secret code work on a different selection of buttons. The breadboard layout that I put together makes it fairly easy to figure out the wiring of the buttons if you understand electronics,

so you might want to use longer interconnecting wires to make the wiring look untidy and to make it more difficult to figure out the secret code.

Summary

In this experiment you learned how two types of logic gates operate and built them into an interesting circuit. Six main logic gates are available to use in electronic circuits: AND, OR, NOT, NAND, NOR, and XOR. The NOT, NAND, NOR, and XOR logic gates are not covered in this book, but you might want to search for them on the Internet and investigate how these logic gates differ from the two described in this chapter. Logic gates are very important in digital circuits; for example, some of the complex ICs inside your home PC will contain logic circuits.

CHAPTER 16

Let There Be Light! Build a Reading Light That Switches On when It Gets Dark

YOU MAY HAVE A SECURITY LIGHT OUTSIDE your house that switches on automatically when it gets dark. This type of light has two main purposes: It provides illumination outside your house to make it more secure at night, and because it is automatic, you don't have to remember to switch the light on and off. In this experiment, shown in Figure 16-1, you'll learn how to make a circuit that acts like this type of security light. It could be used as a small reading light that switches on automatically when it gets dark.

You'll learn about two new components in this experiment—a light-dependent resistor (LDR) and a high-intensity LED. Before you start to build this experiment, it is worth experimenting with a light-dependent resistor to understand how it works. The high-intensity LED is discussed in detail a bit later.

Experimenting with a Light-Dependent Resistor

Think of an LDR as being a bit like the pupil (the black circle) in the center of your eye. Take a look at your eye in a mirror; if you are in a bright room, you'll notice that your eye's pupil shrinks to protect your eye from the bright light. On the other hand, if you are in the dark for a while, your pupil increases in size to let more light in so you can see more clearly in the dark.

The difference between your eye and an LDR is that the *resistance* value of an LDR alters as the light levels change from light to dark. An LDR is ideal for this experiment because it reacts really well to changing light levels.

Before you start the main experiment, you can use your multimeter to understand how changing light levels affect the resistance value of a LDR.

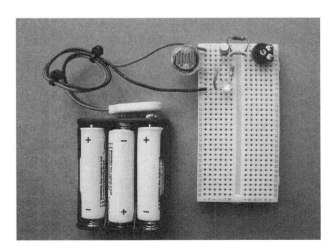

FIGURE 16-1 The reading light

Things You'll Need

The components and equipment that you will need for the first part of the experiment are outlined in the following table. Prepare the items that you need before starting the experiment.

Quantity	Description	Appendix Code
1	Light-dependent resistor (NORPS-12)	12
2	Wire links	—
1	Breadboard	1
—	Multimeter	—

NOTE

The Appendix Code column of the table refers to specific parts that I used in this experiment. Information about sourcing these parts is outlined in the Appendix.

Using the Multimeter with the LDR

Plug the LDR into the piece of breadboard and connect the probes of your multimeter to the wire links, as shown in Figure 16-2.

Switch your multimeter to read resistance. If your multimeter is not auto-ranging, set the resistance level to read the KΩ range; you can adjust the resistance settings as you experiment so that you can get a more accurate resistance reading as you experiment with different light levels.

HINT!

It doesn't matter which way around you connect the multimeter probes, because the resistance is the same whichever way you connect them—try it out and see!

If you are in a fairly bright room near a window or other light source, you should find that the resistance value of the LDR is around 500Ω, as you can see in Figure 16-2. Now slowly wave your hand over the LDR, and the resistance value on your multimeter will change.

You can also experiment with various light levels, either by shining a flashlight onto the LDR or by moving the LDR into a darker area of the room. Use Table 16-1 to record the resistance measurements that you make during your experiments, and then think about the results. How does the resistance value of the LDR correspond to the light levels? For example, does the resistance level increase or decrease as the light levels increase?

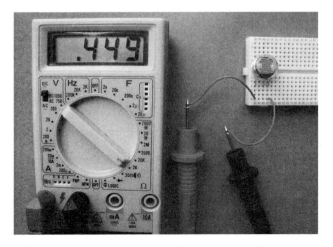

FIGURE 16-2 Connect an LDR to your multimeter.

TABLE 16-1 Record the LDR Resistance Values in This Table

Light Level Applied to the LDR	LDR Resistance Value (Ω)
Near to a window in daylight	466Ω
Center of a room illuminated by daylight	
Hand covering the LDR	
LDR illuminated by a flashlight	
LDR outside in normal daylight	
LDR covered by a piece of cardboard	
LDR in a darkened room	

Experiment 13
Building a Reading Light That Switches On when It Gets Dark

In this experiment, you are going to use a 5mm white LED, which has a much brighter light output than the LEDs that you have come across so far. In fact, this LED is particularly suited to this experiment, because its light output is comparable to a small flashlight bulb and provides a good level of light in the dark. Due to improvements in LED technology over the years, a lot of modern flashlights now use LEDs rather than a filament bulb (aka incandescent lamp).

NOTE

Details of other electronic components are explained in Chapter 3.

The Circuit Diagram

The circuit diagram for the reading light is shown in Figure 16-3. If you have completed the experiment in Chapter 9 and built the temperature sensor, you'll recognize that this circuit also uses a potential divider circuit.

How the Circuit Works

The reading light circuit is made up of four building blocks: power supply, input, control circuitry, and output. This circuit is powered by three 1.5 volt AA batteries that are wired in series to create a 4.5 volt power supply. The light-dependent resistor

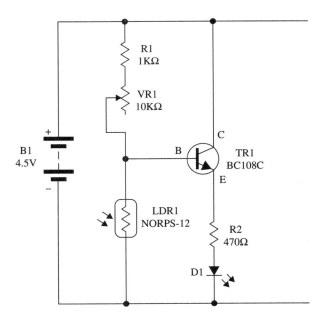

FIGURE 16-3 The circuit diagram for the reading light

(LDR1) is the input component for this project. The variable resistor (VR1) and fixed resistor (R1), in conjunction with the signal from the variable resistor that feeds into the transistor (TR1), form the control part of this project. The high-intensity white LED (D1) and its series resistor (R2) are the output part of this circuit.

The circuit works in a similar way to the temperature sensor in Chapter 9; in fact, if you take a quick look at the circuit diagram for that experiment in Chapter 9, you can see how similar it looks. LDR1 and the resistor network R1 and VR1 form a potential divider circuit, the output voltage of which feeds into the base of the transistor (TR1). The resistance values of resistor VR1 and the LDR have a bearing on the amount of voltage that reaches the base of the transistor. As the base voltage of the transistor increases, this switches on the transistor and then allows current to flow through its collector/emitter junction. This in turn switches on the white LED. The potential divider circuit is configured so that the base of the transistor is switched on as the light levels fall on the LDR. Variable resistor (VR1) acts as a sensitivity control so that you can adjust the trigger point at which the LED illuminates.

Things You'll Need

The components and equipment that you will need for this part of the experiment are outlined in the following table. Prepare the items that you need before starting the experiment.

Code	Quantity	Description	Appendix Code
LDR1	1	Light-dependent resistor (NORPS-12)	12
R1	1	1KΩ 0.5W ±5% tolerance carbon film resistor	—
R2	1	470Ω 0.5W ±5% tolerance carbon film resistor	—
VR1	1	10KΩ 0.2W ±5% variable resistor	—
D1	1	5mm high-intensity white LED (6000mcd)	7
TR1	1	BC108C NPN transistor	9
B1	1	4.5V battery holder	15
B1	3	1.5V AA batteries	—
B1	1	PP3 battery clip	17
—	1	Breadboard	1
—	—	Wire links	—

NOTE

The Appendix Code column of the table refers to specific parts that I used in this experiment. Information about sourcing these parts is outlined in the Appendix.

If you carried out the experiment with the LDR at the beginning of this chapter, you'll realize that the resistance value of the LDR increases as the light levels reduce and the resistance value decreases as the light level increases. This changing resistance causes the LED to switch on and off as the light levels change.

The Breadboard Layout

The breadboard layout for the reading light experiment is shown in Figure 16-4, which also identifies the component codes to help you to identify each component. Take care when inserting the transistor (TR1) to make sure that the pins are wired in the correct way. Also make sure that the LED is inserted correctly so that the large electrode inside the LED (the cathode) is connected to the negative (–) side of the battery.

 **HINT!**

Chapter 3 shows you how to identify the pin-outs of the various components used in this experiment.

FIGURE 16-5 Close-up of the wiring around the LDR and variable resistor

A close-up photograph of the experiment is shown in Figure 16-5 to give you a better view of some of the pin connections of the components.

Once you have built the breadboard layout, it is time to experiment!

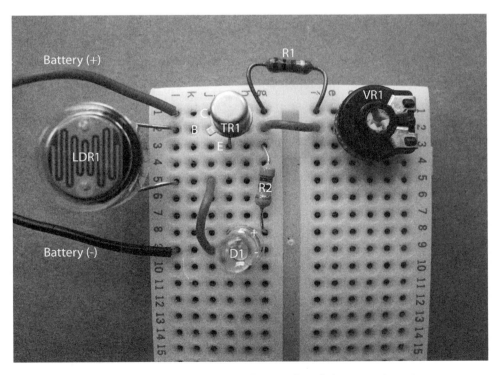

FIGURE 16-4 The breadboard layout for the reading light experiment

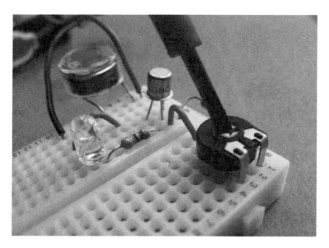

FIGURE 16-6 Wind the variable resistor so that it is fully clockwise until it stops.

Time to Experiment!

First, make sure that you are working in a well-lit room. Before you connect the battery to the circuit, you need to adjust the sensitivity setting of the circuit by using a small screwdriver to turn the variable resistor screw fully clockwise until it stops, as shown in Figure 16-6.

Now apply the battery to the circuit, and the LED will switch off or will be illuminated slightly. If the LED is illuminated, use the screwdriver to slowly wind the variable resistor counterclockwise until the LED switches off. If the LED is off, you don't need to adjust the variable resistor at the moment.

Slowly move your hand over the LDR so that it creates a shadow; the LED will illuminate. The darker the area around the LDR, the brighter the LED becomes. Try using a piece of cardboard to cover the LDR to activate the LED.

HINT!

If the LED switches on at the slightest shadow, try turning the variable resistor a bit further counterclockwise to make the circuit less sensitive.

Experiment with turning the variable resistor screw in various positions to see what effect this has on the sensitivity of the circuit. You should find a "sweet spot" where the LED remains switched off until the surroundings become very dark.

NOTE

The "sweet spot" for my sensor was when the variable resistor screw was set at roughly the halfway position; depending on component tolerances you may find that your circuit differs slightly.

Once you are happy with the settings, take the breadboard into a dark room and switch the light on; the LED should light up in the dark and switch off when it's light. Now turn the light off so that the room is dark again; the LED will switch on almost immediately.

You have now created a circuit that illuminates an LED whenever it gets dark. You can probably think of a number of different applications for this circuit—it could be used as an reading light in your bedroom that automatically lights when it gets dark.

Summary

In this chapter, you learned how to build a potential divider circuit using an LDR, which can provide a variable voltage output depending on the amount of ambient light that shines onto it. You also learned that an ultra-bright LED can be used to generate a light output bright enough to create a small reading light.

 INTERESTING FACT

You probably noticed that the transistor action of this experiment and the temperature sensor in Chapter 9 caused the LED to switch on or off gradually. Sometimes you might want to produce a circuit design with a "clean" switching action instead, so that the LED switches on or off without fading, just like when you flick a light switch on or off. You can achieve this by using a *Schmitt trigger* circuit that produces the desired effect. This book does not experiment with Schmitt trigger circuits, but as your electronics knowledge grows, you might find it fun to work with this type of circuit.

PART FIVE

Electronic Game Experiments

CHAPTER 17

Pick a Number! Create a Random Number Generator

GAMES ARE PART OF OUR EVERYDAY LIVES, whether they are simple board games, more complex handheld electronic games, or games that use consoles. We like to play games because they can be exciting and they help us to pass the time. I'm sure that lots of times you have been so engrossed in playing a game that you lose track of time—before you know it, several hours have passed. This part of the book contains experiments that let you explore how you can use various electronic building blocks to creating your own electronic game circuits.

We'll start with an experiment to create a random number generator, kind of like an electronic die. This experiment introduces you to some new electronic components and shows you how to build an electronic random number generator that produces a random number between 0 and 9.

Experiment 14
Creating a Random Number Generator

If you have played a board game, you will have noticed that they usually require at least one die (or two dice), like the one in Figure 17-1.

A die has six sides and contains a number of dots that represent the numbers 1 to 6. To use a die, you

FIGURE 17-1 A simple die

usually throw it on a tabletop, and the number that lands on top tells you to move a certain number of spaces in the game.

This experiment is split into two parts: the first part introduces you to a new LED component (a seven-segment display), and the second part shows you how to build the random number generator.

  BE CAREFUL!

This experiment flashes LEDs on and off. If you suffer from epilepsy or are affected by flashing lights, this experiment is not for you.

Experimenting with a Seven-Segment LED Display

For this experiment, you will be using an LED display that can show numbers between 0 and 9. This type of LED is called a *seven-segment display* and is shown in Figure 17-2. (This display contains seven individual LED segments configured in the shape of a number *8*, plus an additional LED for the decimal point, which means that the display actually contains eight LEDs! You'll ignore the decimal point in this experiment, however, and concentrate on the seven main segments.) This display might look familiar, because similar displays are used in lots of different household applications, such as the timer on a microwave and the lighted display on a DVD player.

Each LED can be illuminated individually, and the combination of LEDs illuminated will determine which number is displayed. Each LED segment is identified by a letter *A* to *G*; these are not shown on the seven-segment display but are usually shown on circuit diagrams and manufacturers datasheets.

TABLE 17-1 How Numbers Are Created on a Seven-Segment Display

Number Displayed	A	B	C	D	E	F	G
0	On	On	On	On	On	On	Off
1	Off	On	On	Off	Off	Off	Off
2	On	On	Off	On	On	Off	On
3	On	On	On	On	Off	Off	On
4	Off	On	On	Off	Off	On	On
5	On	Off	On	On	Off	On	On
6	On	Off	On	On	On	On	On
7	On	On	On	Off	Off	Off	Off
8	On	On	On	On	On	On	On
9	On	On	On	On	Off	On	On

Table 17-1 shows which LED segments must be switched on to produce the numbers *0* to *9*. It's like reading a secret code—each number has its own special combination. If, for example, you want to show a number *3* on the display, the LED would need to illuminate segments A, B, C, D, and G. Notice that to create a number *8*, all segments are illuminated.

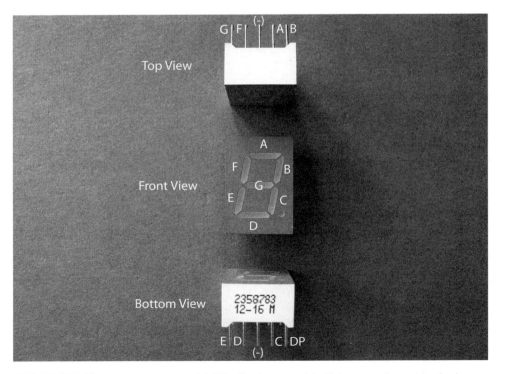

FIGURE 17-2 The seven-segment LED display used in this experiment includes the pin connections on the top and the bottom of the component. (Note that if you use a different seven-segment display in your experiment, the pin-outs may be different.)

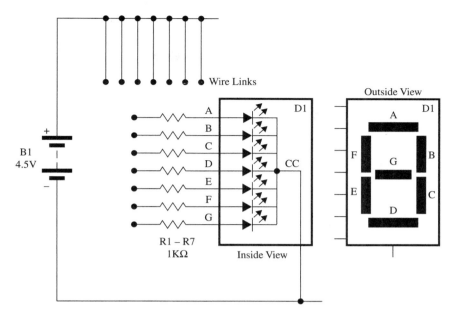

FIGURE 17-3 The circuit diagram for the seven-segment display experiment

The type of seven-segment display that you are going to use in this experiment is a *common cathode (CC)* type, which means that all of the cathode leads (negative connections) of each of the LEDs are linked together, like the one shown in the circuit diagram (Figure 17-3) for this part of the experiment.

 **NOTE**

Another version of seven-segment LED display is also available, called a *common anode* type. Can you guess how each of the LED segments is connected in this type of display?

Things You'll Need

The parts that you'll require for this part of the experiment are outlined in the following parts list. Prepare the items that you need before starting the experiment.

Code	Quantity	Description	Appendix Code
D1	1	Seven-segment CC display (red)	8
R1–R7	7	1KΩ 0.5W ±5% tolerance carbon film resistor	—
B1	1	4.5V battery holder	15
B1	3	1.5V AA batteries	—
B1	1	PP3 battery clip	17
—	1	Breadboard	1
—	—	Wire links	—

 **NOTE**

The Appendix Code column of the table refers to specific parts that I used in this experiment. Information about sourcing these parts is outlined in the Appendix.

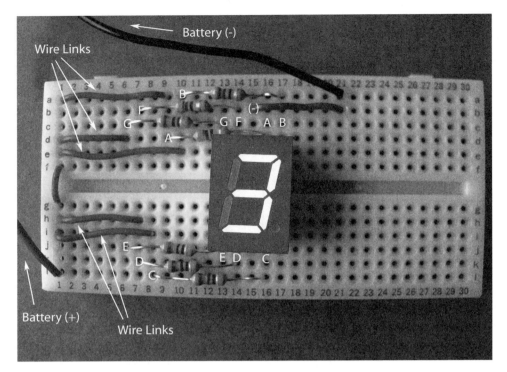

FIGURE 17-4 Experimenting with a seven-segment display to create a number *3*

You can experiment with the CC display using seven individual 1KΩ resistors and a piece of breadboard, as shown in Figure 17-4. In the figure, I have used five of the wire links to connect various resistors to the positive side of the battery to create the number *3* on the display. You can see that these five wire links are connected to resistors A, B, G, C, and D.

Once you have built the breadboard layout to create the number *3*, you can experiment a bit more by adding or removing wire links to illuminate various

LED combinations. Use Table 17-1 as a guide to creating the numbers *0* to *9* on the display. Then try different combinations to see if you can produce other shapes and even letters on the display. You can use Table 17-2 to record some of your findings; an example is shown at the top of the table.

INTERESTING FACT

There are a total of 128 different display combinations that can be produced using a seven-segment display. This is because each segment has one of two states; it is either illuminated or switched off. The total number of display combinations can be calculated by multiplying 2 by itself 7 times; $2 \times 2 \times 2 \times 2 \times 2 \times 2 \times 2 = 128$. See if you can create them all!

Building a Random Number Generator

Now that you've experimented with a seven-segment display and understand how it works, you're ready to build a random number generator circuit.

TABLE 17-2 What Other Shapes Can You Create?

Shape Displayed	A	B	C	D	E	F	G
Letter *A*	On	On	On	Off	On	On	On

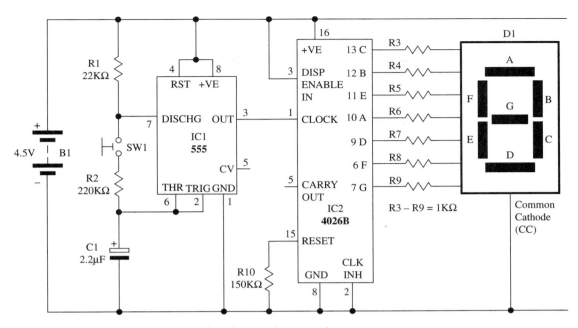

FIGURE 17-5 The circuit diagram for the random number generator

The Circuit Diagram

If you have been reading this book from the start and have been working through each of the experiments, you should be familiar with circuit diagrams. The circuit diagram for the random number generator experiment is shown in Figure 17-5. Do you recognize parts of this circuit? It might look complicated, but if you break it down into its individual building blocks, it is much easier to understand.

How the Circuit Works

The random number generator is made up of four building blocks: power supply, input, control circuitry, and output. This circuit is powered by three 1.5 volt AA batteries that are wired in series to create a 4.5 volt power supply. The normally open switch (SW1) creates the input for this circuit. There are two main parts to the control circuit:

- A 555 timer (IC1) that is wired in astable mode; this creates the clock part of the circuit.

- A 4026B seven-segment counter chip (IC2), which is a new component to you. It counts the clock signals from the 555 timer and converts

them into a code that can drive the seven-segment display.

The seven-segment LED display (D1) and its series resistors (R3–R9) create the output part of this circuit.

Assume for the moment that push-button switch SW1 is being pressed to close the circuit. When the battery is connected to the circuit, the 555 timer IC (IC1), which is wired as an astable circuit, starts to produce a train of pulses on its output pin 3. The timing speed of the astable circuit is determined by resistors R1 and R2 and capacitor C1. The output pin of the 555 timer is connected to the "clock" pin of IC2, which is a seven-segment counter IC. Every time IC2 is "clocked," it produces a coded output on its seven pins 6, 7, 9, 10, 11, 12, and 13, which produce the numbers *0* to *9* when connected to the correct LED segments in the display. These coded outputs are similar to the codes in Table 17-1. If you remove your finger from the switch (SW1), this takes resistor R2 out of the circuit and stops the 555 timer from producing pulses from pin 3. You'll see how this works shortly.

Things You'll Need

The components and equipment that you will need for this part of the experiment are outlined in the following parts list. Prepare the items that you need before starting the experiment.

Code	Quantity	Description	Appendix Code
IC1	1	555 timer IC	18
IC2	1	4026B seven-segment counter	19
R1	1	22KΩ 0.5W ±5% tolerance carbon film resistor	—
R2	1	220KΩ 0.5W ±5% tolerance carbon film resistor	—
R3–R9	7	1KΩ 0.5W ±5% tolerance carbon film resistor	—
R10	1	150KΩ 0.5W ±5% tolerance carbon film resistor	—
C1	1	2.2µF electrolytic capacitor *and* a 10nF ceramic disk capacitor (minimum 10 volt rated)	—
D1	1	Seven-segment common cathode display (red)	8
SW1	1	Normally open switch	24
B1	1	4.5V battery holder	15
B1	3	1.5V AA batteries	—
B1	1	PP3 battery clip	17
—	1	Breadboard	2
—	—	Wire links	—

NOTE

The Appendix Code column of the table refers to specific parts that I used in this experiment. Information about sourcing these parts is outlined in the Appendix.

The Breadboard Layout

NOTE

Refer to Chapter 3 for building breadboard layouts and fault-finding guidelines.

The breadboard layout for this experiment is shown in Figure 17-6. Follow this layout to build your breadboard—take your time, because many connections are required. Make sure that you use the 2.2µF electrolytic capacitor for C1 when you first build the breadboard layout, ensuring that the positive lead of the capacitor is connected to pin 2 of the 555 timer.

To help you better see the circuit, the display connections are shown in more detail with the seven-segment display removed in Figure 17-7.

Figure 17-8 shows the layout from a different angle to help you identify the connections around the 555 timer and switch SW1.

Time to Experiment!

Once you have built the breadboard layout, connect the 4.5 volt battery to the circuit. The display shows a number—either a *0* or a *1*. Now press the switch and keep your finger on it; the display starts to count slowly from *0* to *9* in a never-ending loop. If you remove your finger from the switch, the count

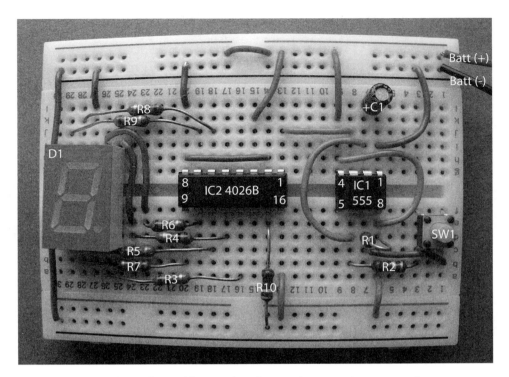

FIGURE 17-6 The breadboard layout for the random number generator

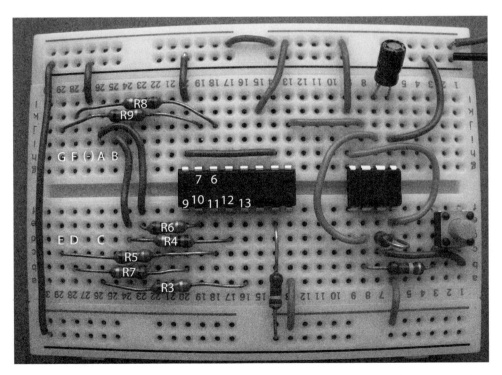

FIGURE 17-7 The breadboard layout with the seven-segment display removed

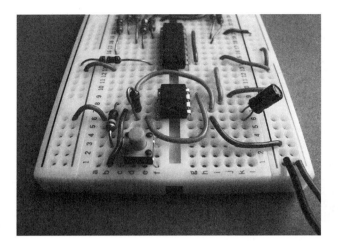

FIGURE 17-8 Wiring around the 555 timer and switch

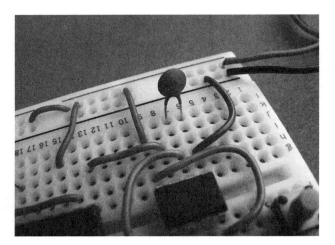

FIGURE 17-9 Change the capacitor to a 10nF ceramic disk version.

stops and remains on a number until you press the button again. Impressive, isn't it!

You might be thinking at this point that the circuit is not really suited to creating a random number (as a die can do when you toss it), and you are right; you need to make a few adjustments to get this right.

First, increase the speed of the 555 timer to make the count go a lot faster. Do you know how you might do this? If you read Chapter 5, you'll remember that you can change the speed of a 555 timer by altering the values of its timing resistors and capacitors. In this experiment, you'll change the value of capacitor C1 to do this.

Remove the battery from the circuit and remove the 2.2μF capacitor (C1). Replace it with a 10nF ceramic disk type, as shown in Figure 17-9.

Next, connect the battery again to see a number on the display. Then press the switch for a few seconds, and you should see a flashing number *8* on the display. This happens because the 555 timer is producing clock pulses at a much faster rate because of the capacitor value; the display is changing from *0* to *9* very quickly, making it difficult to see which number is being displayed.

Remove your finger from the button, and you'll see a number on the display. Since you can't see which

number will appear when you remove your finger (because the display is changing so quickly), the number that appears is equal to a random number. Now press and release the button a few times, and you will see that it is very difficult to predict which number you will end up with.

Congratulations! You have built an electronic random number generator!

Summary

In this chapter, you learned about seven-segment displays and how they can be used to create numbers 0 to 9 along with various letters and shapes. You also learned how two different electronic circuits can be joined together to produce an interesting effect.

If you're in a mathematical mood, you could calculate the frequency of the 555 timer by using the astable formula from Chapter 5 to see how the astable speed changes when you alter the value of C1 in your experiment.

Keep the breadboard layout from this experiment, because you will be making some modifications to it in Experiment 15 in Chapter 18.

CHAPTER 18

Heads or Tails? Flip an Electronic Coin

FOR SOME GAMES, YOU FLIP A COIN to see which player starts first. Each player shouts "Heads!" or "Tails!" to predict which side of the coin will face up before the coin is flipped—each player has a 50/50 chance of winning the coin toss. But a coin has no electronics inside! As an inventor, you want to find a way of creating an electronic circuit that does the same thing. This experiment shows you how to build it.

If you haven't built the random number generator experiment in Chapter 17, you should do so before starting this experiment. The electronic coin uses a lot of the core electronic circuitry of the random number generator and shows you how, with some creative thinking, you can modify the circuit to produce a slightly different end result.

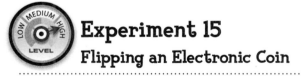

Experiment 15
Flipping an Electronic Coin

In this experiment you'll use a seven-segment display to create a circuit that simulates a coin toss to show heads or tails. Before you read any further, though, think about how you might use some of the circuit modules that you have learned about in this book to create such a circuit. You might already have some specific ideas about how to make it; if

that's the case, that's great, because you are already starting to think like an inventor.

You could create an electronic coin in many different ways. In this experiment, you'll see the circuit that I created. I'll explain my thought process behind the circuit design as I go.

BE CAREFUL!

This experiment flashes LEDs on and off. If you suffer from epilepsy or are affected by flashing lights, this experiment is not for you.

Making the Display Show Heads or Tails

If you have already built the random number generator, you've experimented with a seven-segment display and you discovered that you can create many different numbers, letters, and shapes on the display, depending on how you illuminate various LED segments. Figure 18-1 shows you how I imagined a seven-segment display would look for *this* experiment.

Of course, a single seven-segment display is unable to show the words "heads" and "tails," so I've shortened these to a letter *h* and a letter *t* to show the results of the coin flip. If you permanently illuminate segments E, F, and G of the display, it

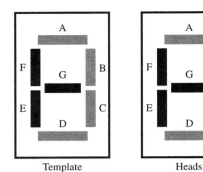
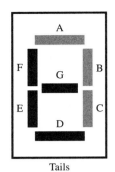

FIGURE 18-1 The three versions of the seven-segment display

provides a good *template* upon which to create the two letters. Along with the template, then, you can simply illuminate segment C to produce the letter *h* or segment D to produce the letter *t*.

Now that the display is taken care of, you need a circuit so that segments C and D can be illuminated alternately to create the two letters.

Decade Counter

For the random number generator experiment, you used a 4026B IC to produce the output codes to show the numbers *0* to *9* on the display. For this circuit, you'll use a different IC called a *74HC4017 decade counter*, which has ten output pins (Q0 to Q9).

It is called a decade counter because it is able to count to ten. Every time this IC receives a *clock signal* (for example, from the output of a 555 astable timer IC), each output pin activates in turn, as shown in Table 18-1.

You'll soon see how you can use just three of these output pins to create your electronic coin.

The Circuit Diagram

The complete circuit diagram is shown in Figure 18-2.

If you compare this circuit diagram with the diagram for the random number generator, you'll see quite a few similarities.

TABLE 18-1 How the Outputs of a 74HC4017 IC Work

Clock Pulse	Q0	Q1	Q2	Q3	Q4	Q5	Q6	Q7	Q8	Q9
1	On	Off	Off	Off	Off	Off	Off	Off	Off	Off
2	Off	On	Off	Off	Off	Off	Off	Off	Off	Off
3	Off	Off	On	Off	Off	Off	Off	Off	Off	Off
4	Off	Off	Off	On	Off	Off	Off	Off	Off	Off
5	Off	Off	Off	Off	On	Off	Off	Off	Off	Off
6	Off	Off	Off	Off	Off	On	Off	Off	Off	Off
7	Off	Off	Off	Off	Off	Off	On	Off	Off	Off
8	Off	Off	Off	Off	Off	Off	Off	On	Off	Off
9	Off	Off	Off	Off	Off	Off	Off	Off	On	Off
10	Off	Off	Off	Off	Off	Off	Off	Off	Off	On
11	On	Off	Off	Off	Off	Off	Off	Off	Off	Off
12	Off	On	Off	Off	Off	Off	Off	Off	Off	Off

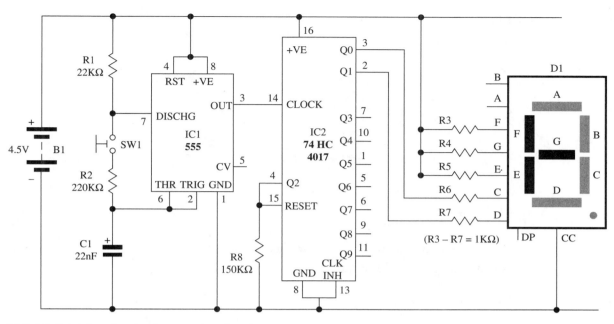

FIGURE 18-2 The circuit diagram for the electronic coin

How the Circuit Works

The electronic coin is made up of four building blocks: power supply, input, control circuitry, and output. This circuit is powered by three 1.5 volt AA batteries that are wired in series to create a 4.5 volt power supply. The normally open switch (SW1) creates the input for this circuit. There are two main parts to the control circuit:

- A 555 timer (IC1) that is wired in astable mode; this creates the clock part of the circuit

- A 74HC4017 decade counter chip (IC2), which counts the clock signals from the 555 timer and converts them into the code shown in Table 18-1

The seven-segment LED display (D1) and its series resistors (R3–R7) create the output part of this circuit.

Notice that the seven-segment display (D1) has three of its LED segments permanently illuminated to create the template shown in Figure 18-1. These segments are E, F, and G, and they are connected to the positive supply rail using three resistors—R3, R4, and R5—so that they are permanently illuminated.

Assume for the moment that push-button switch SW1 is being pressed to close the circuit. When the battery is connected to the circuit, the 555 timer (IC1), which is wired as an astable circuit, starts to produce a fast train of pulses on its output pin 3. The timing speed of the astable circuit is determined by resistors R1 and R2 and capacitor C1. The output pin of the 555 timer is connected to the clock pin of IC2, which is a 74HC4017 decade counter IC. Only three output pins of IC2 are being used: pin 2 is used to activate segment D of the seven-segment display, pin 3 is used to activate segment C, and pin 4 is connected to the reset pin 15 of IC2 to reset the counter when it is activated. This means that IC2 operates as shown in Table 18-2 every time a clock pulse is received, and this continues in a never-ending loop.

If you remove your finger from the switch (SW1), this takes resistor R2 out of the circuit and stops the 555 timer from producing pulses at pin 3; this shows the result of the coin toss to appear on the display.

TABLE 18-2 How IC2 Drives the Display

Clock Pulse	Q0 (pin 3)	Q1 (pin 2)	Q2 (pin 4)	Display Shows
1	*On*	Off	Off	h (heads)
2	Off	*On*	Off	t (tails)
3	Off	Off	*On*	Resets the counter
4	*On*	Off	Off	h (heads)
5	Off	*On*	Off	t (tails)
6	Off	Off	*On*	Resets the counter

Things You'll Need

The components and equipment that you will need for this experiment are outlined in the following parts list. Prepare the items that you need before starting the experiment.

Code	Quantity	Description	Appendix Code
IC1	1	555 timer IC	18
IC2	1	74HC4017 decade counter IC	20
R1	1	22KΩ 0.5W ±5% tolerance carbon film resistor	—
R2	1	220KΩ 0.5W ±5% tolerance carbon film resistor	—
R3–R7	5	1KΩ 0.5W ±5% tolerance carbon film resistor	—
R8	1	150KΩ 0.5W ±5% tolerance carbon film resistor	—
C1	1	22nF ceramic disk capacitor (minimum 10 volt rated)	—
D1	1	Seven-segment common cathode display (red)	8
SW1	1	Normally open switch	24
B1	1	4.5V battery holder	15
B1	3	1.5V AA batteries	—
B1	1	PP3 battery clip	17
—	1	Breadboard	2
—	—	Wire links	—

NOTE

The Appendix Code column of the table refers to specific parts that I used in this experiment. Information about sourcing these parts is outlined in the Appendix.

The Breadboard Layout

NOTE

Refer to Chapter 3 for building breadboard layouts and fault-finding guidelines.

The breadboard layout for this experiment is shown in Figure 18-3.

If you have already built the random number generator, you'll find it easy to modify that breadboard layout to create this one. You're

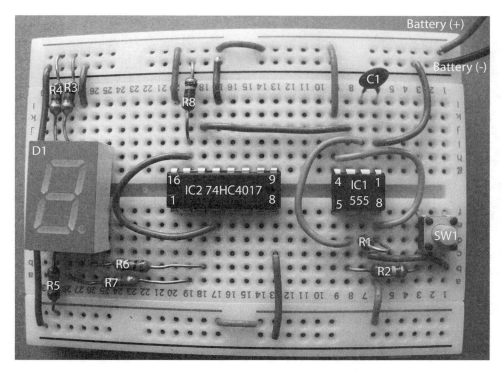

FIGURE 18-3 The breadboard layout for the electronic coin

mainly replacing IC2 and changing some of the connections around the display. Notice the orientation of IC2 in this experiment: it differs from the orientation of IC2 in the last experiment.

Figure 18-4 shows a close-up of the display connections with the display removed. Figure 18-5 shows a close-up of the wiring around IC1 and SW1.

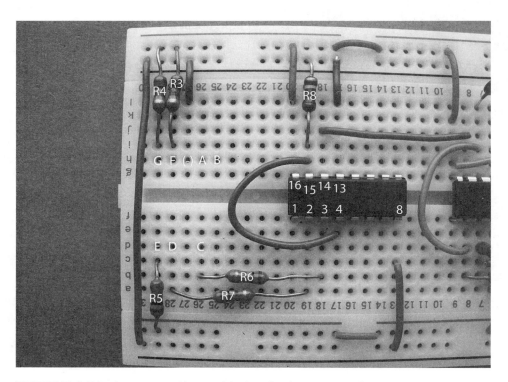

FIGURE 18-4 Display connections with the display removed

FIGURE 18-5 Close-up of the wiring around the 555 timer IC

Time to Experiment!

Now that you have built the breadboard layout, it's time to play. Connect the battery to the circuit, and the display shows either a letter *h* (heads) or a letter *t* (tails), as shown in Figure 18-6.

FIGURE 18-6 Displays showing "heads" and "tails"

Press the push-button switch (SW1) for a few seconds; you're flipping the electronic coin! The display shows a letter *b*, because segments C and D are switching on and off very quickly, so it looks like both are illuminated at the same time. Now remove your finger from the button and you should see an *h* or a *t* on the display. Try flipping the coin a few more times to see if you can predict whether the coin will land on heads or tails.

Try changing the value of C1 to see what effect speeding up or slowing down the speed of the 555 timer has on the display. And if you decide to experiment with an electrolytic capacitor, make sure that its negative lead is connected to pin 1 of the 555 timer.

Summary

This experiment showed how I designed this electronic coin. It demonstrates how easy it is to connect various electronic circuits to produce a fairly complex design. Keep this breadboard layout when you are finished experimenting, because you will be using it again when you build the final experiment in Chapter 20.

CHAPTER 19

Ready, Aim, Fire! Get Ready for Infrared Target Practice

WHEN I WAS A KID, I ALWAYS DREAMED ABOUT owning a laser gun—the type used in sci-fi movies to shoot those nasty aliens from faraway planets. This experiment doesn't show you how to make a laser gun, but it does show you how you can shoot an electronic target using an invisible beam of infrared light. The experiment, shown in Figure 19-1, shows you how to make an infrared target that flashes an LED whenever it is hit by an infrared beam of light.

Experiment 16
Getting Ready for Infrared Target Practice

Before you start infrared target practice, you need to build an infrared gun. In this experiment, you'll be using a common household infrared remote control as your "laser gun." You probably have some old discarded remotes that you no longer use at your house. Ask your parents if they still have an old remote for a TV or video recorder that is no longer used or working.

BE CAREFUL!

This experiment flashes an LED on and off. If you suffer from epilepsy or are affected by flashing lights, this experiment is not for you.

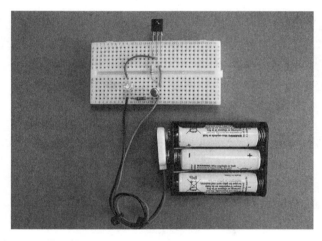

FIGURE 19-1 The infrared target receiver circuit

Every infrared remote contains an infrared LED that sends out an invisible beam of light. Whenever you press the volume button on your TV remote, for example, the circuitry inside sends out a special coded signal that flashes the infrared LED on and off in a specific way. If you look at the end of a remote (the end that you point at the TV to turn up the volume or change a channel), you might be able to see an LED, but when you press any button on your remote you won't be able to see it light up. It is possible to "see" this infrared light, but only if you view the LED through a digital camera. This is because the sensor inside a digital camera is able to pick up infrared light. Try this out by pressing the

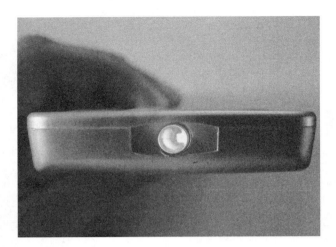

FIGURE 19-2 The infrared LED cannot be seen by the human eye, but it can be seen by a digital camera.

volume button on your remote control and viewing the infrared LED through a digital camera.

Figure 19-2 shows the infrared LED on my remote control as it's illuminated.

Every TV or DVD player has a special infrared receiver circuit built into it that receives and decodes the infrared light signal from a remote and increases the volume or changes a channel, for example. You can use a number of circuit methods to receive infrared light. This experiment

uses a special component that contains an infrared photodiode built into it; it can pick up and decode infrared signals at more than 10 feet away.

 INTERESTING FACT

Before infrared remote controls for your TV or video player were invented you had to get up and walk to your TV to change the channel! Some early remote controls used a long piece of cable to connect your video player to the remote control, and some TV remotes used sound (usually a clicking noise) to change the channel.

The Circuit Diagram

The circuit diagram for the infrared target practice receiver is shown in Figure 19-3.

You can see that the circuit is very simple, with only five components required. This is because the infrared sensor and most of the circuitry needed to receive and amplify an infrared signal are provided by the single sensor component (IC1). This component has only three leads—a bit like a

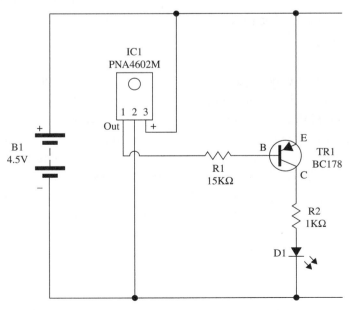

FIGURE 19-3 The circuit diagram for the target practice receiver

transistor—but the three leads for this device have the following functions:

- **Pin 1** Output signal
- **Pin 2** Battery negative
- **Pin 3** Battery positive

The operation of this component is described in more detail shortly.

How the Circuit Works

The electronic target practice experiment is made up of four building blocks: power supply, input, control circuitry, and output. The circuit is powered by three 1.5 volt AA batteries that are wired in series to create a 4.5 volt power supply. The infrared receiver (IC1) creates the input part of this circuit. The PNP transistor (TR1) and its base resistor (R1) form the control part of this circuit. The high intensity white LED (D1) and its series resistor (R2) create the output part of this circuit.

The circuit is fairly straightforward. When the circuit is connected to the battery, the output pin 1

of the infrared receiver (IC1) is a positive 4.5 volts. This pin is connected to the base of the PNP transistor (TR1) via a series resistor (R1), which means that the transistor is switched off.

NOTE

A PNP transistor requires a negative signal on its base connection to switch it on, unlike an NPN transistor, which requires a positive voltage. You will also notice that the emitter of the PNP transistor is connected to the positive battery connection, unlike an NPN version, which normally has its emitter connected to the negative side of the battery.

Because the transistor is switched off, current is unable to flow through the emitter/collector junction, and this means that the LED (D1) is switched off. Now for the clever bit. If you press any button on your remote control, the infrared signal that it sends out is received by the infrared

Things You'll Need

The components and equipment that you will need for this experiment are outlined in the following parts list. Prepare the items that you need before starting the experiment.

Code	Quantity	Description	Appendix Code
IC1	1	Infrared photodiode and amplifier (PNA4602M)	13
TR1	1	BC178 PNP transistor	10
R1	1	15KΩ 0.5W ±5% tolerance carbon film resistor	—
R2	1	1KΩ 0.5W ±5% tolerance carbon film resistor	—
D1	1	5mm high-intensity white LED (6000mcd)	7
B1	1	4.5V battery holder	15
B1	3	1.5V AA batteries	—
B1	1	PP3 battery clip	17
—	1	Breadboard	1
—	—	Wire links	—

NOTE

The Appendix Code column of the table refers to specific parts that I used in this experiment. Information about sourcing these parts is outlined in the Appendix.

sensor (IC1), and each of the coded pulses received produces a negative signal on output pin 1. Whenever this happens, the negative pulses activate the base of the PNP transistor (TR1), and this briefly switches the LED (D1) on. This has the effect of flashing the LED (D1) on and off in synchronization with the coded pulses that are sent by the remote control. When you stop pressing a button on the remote, the circuit returns back to its ambient state and the LED is switched off again.

The Breadboard Layout

NOTE

Refer to Chapter 3 for building breadboard layouts and fault-finding guidelines.

The breadboard layout for the infrared receiver is shown in Figure 19-4. This will help to identify the component codes that are shown on the circuit diagram and in the parts list.

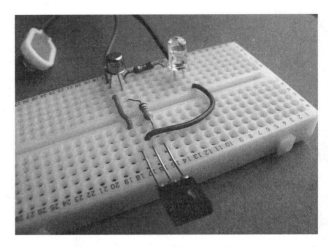

FIGURE 19-5 A close-up of the component layout; notice the orientation of the transistor and the LED. You can also see the infrared photodiode that is built into the sensor component.

Figure 19-5 shows the breadboard layout from a different angle so that you can see the pin connections of the transistor and the LED in more detail.

Once you have built the breadboard layout and checked it over carefully, it's time to experiment!

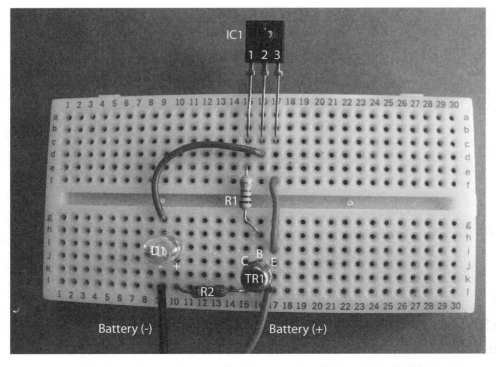

FIGURE 19-4 The breadboard layout for the target practice experiment

Time to Experiment!

Connect the 4.5 volt battery to the breadboard layout, and the LED flashes briefly and then switches off again; don't worry, because this is normal. Once the circuit is powered up, find yourself an infrared remote and point it at the sensor (IC1). Then press any of the buttons on the remote and the LED (D1) will flash on and off brightly. If you keep your finger on the button, the LED will continue to flash on and off, as shown in Figure 19-6.

Try pressing different buttons on your remote control and see what effect this has on the speed of the LED flashes. You can also experiment to see how far away you can move from the infrared target circuit and still illuminate the LED.

You should find that the infrared sensor component is quite sensitive and that it may be possible to trigger the LED even when you are pointing the remote control away from the sensor. This is because the sensor is picking up the infrared light signals that are bouncing off walls and furniture.

Clearly, you need to modify the infrared target receiver slightly to make it less sensitive and to make it a proper target. One method that I used is to find some black insulation tape—the type that

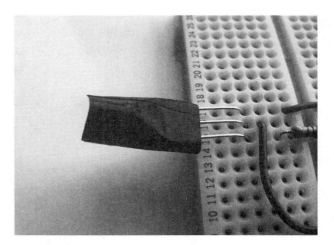

FIGURE 19-7 Wrap some black tape around the sensor to make it less sensitive.

electricians use—and carefully wrap a small piece around the infrared sensor, as shown in Figure 19-7.

You might need to experiment with the amount of tape and the orientation of the sensor, and you should eventually be able to create an infrared sensor that will activate the LED only when the remote is pointed directly at it. At this point, you'll know that you have made a fairly accurate infrared target. You and your friends could take turns shooting the target, and whoever makes the LED flash the most wins the game.

Summary

In this experiment, you discovered that integrated circuits don't always follow the normal 8- or 16-pin convention, and that they come in various shapes and sizes and can perform many different functions. You could have built the infrared-sensing part of this circuit using a simple infrared photodiode, but you would have also required additional components to amplify the signal, and this would have made the circuit larger and slightly more complicated. The use of the IC simplified the circuit a lot. You also met the PNP transistor for the first time in this experiment, and you learned how its connections and operation differ slightly from those of the NPN transistors that you have already used.

FIGURE 19-6 Pressing any button on the remote control causes the LED to flash on and off.

CHAPTER 20

Let's Make Some Noise! Build a Sound Effects Generator

SADLY, YOU HAVE REACHED THE LAST EXPERIMENT of the book. But don't worry, because it's a noisy one! You've probably played many electronic games that produce sound effects or music to add atmosphere. In this final experiment, you'll discover how three different circuit modules can be connected together to produce a device that creates some interesting noises. The complete sound effects generator is shown in Figure 20-1.

 Experiment 17
Building a Sound Effects Generator

The circuit in this experiment allows you to try out various components to create a combination of four different sound frequencies to produce various sound effects. Each sound frequency is activated one at a time. This type of circuit is also known as a *sequencer*. This experiment also demonstrates how you can easily attach various electronic circuits together to produce a more complex circuit.

This experiment uses three circuits that you should have worked with already:

- 555 astable timer circuit (Chapter 5)
- 74HC4017 decade counter circuit (Chapter 18)
- 555 astable sound generator (Chapter 10)

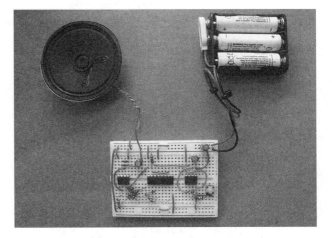

FIGURE 20-1 The sound effects generator

 NOTE

Before you build this experiment, you need to build the electronic coin experiment from Chapter 18, because the operation of that circuit has some relevance to this one, and the breadboard layout for the electronic coin experiment can be easily modified to create the sound effects generator.

The Circuit Diagram

The circuit diagram for this experiment is shown in Figure 20-2. If you have completed most of the experiments in this book, you should be able to

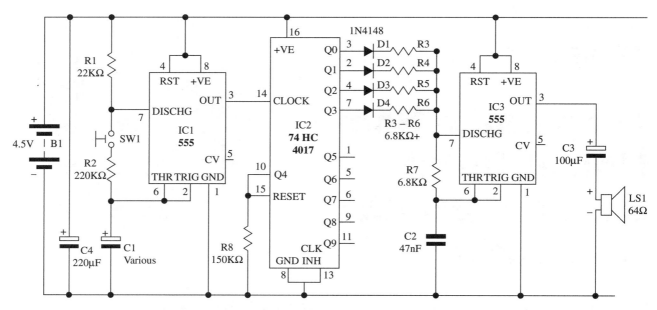

FIGURE 20-2 The circuit diagram for the sound effects generator

identify all of the components by now; you might even recognize how the circuit operates.

How the Circuit Works

The sound effects generator is made up of four building blocks: power supply, input, control circuitry, and output. This circuit is powered by three 1.5 volt AA batteries that are wired in series to create a 4.5 volt power supply. The normally open switch (SW1) creates the input for this circuit. There are three main parts to the control circuit:

1. A 555 timer (IC1) that is wired in astable mode; this creates the clock part of the circuit.

2. A 74HC4017 decade counter chip (IC2) that counts the clock signals from the 555 timer and converts them into a four-digit sequencer code.

3. A 555 timer (IC3) that is wired in astable mode and is connected to a speaker to create the sounds.

The speaker (LS1) and capacitor (C3) create the output part of this circuit.

Assume for the moment that push-button switch (SW1) is being pressed to close the circuit. When the battery is connected to the circuit, the 555 timer

(IC1), which is wired as an astable circuit, starts to produce a fast train of pulses on its output pin 3. The timing speed of the astable circuit is determined by resistors R1 and R2 and capacitor C1. The output pin of the 555 timer is connected to the "clock" pin of IC2, which is a 74HC4017 decade counter IC. Four output pins of IC2 are being used (pins 2, 3, 4, and 7), and each of these is connected to a series resistor (R3 to R6) via individual diodes (D1 to D4). Every time a clock pulse is received, each output switches on and then off in turn. Once pin 10 is activated, it resets IC2 by activating pin 15, and the count starts again. This creates a "count-by-four counter," and the four outputs operate in a never-ending loop, as shown in Table 20-1.

Each time pins 3, 2, 4, and 7 of IC2 are activated, this adds a resistor into the third part of the circuit, which is the 555 timer (IC3) that is used to create the sounds.

The 555 timer (IC3) is also configured in astable mode like IC1, but the output pin of this part of the circuit is connected to a speaker via capacitor C3 to produce a noise output.

And here's the clever bit: Changing the value of the resistors (R3 to R6) causes the 555 astable timer

TABLE 20-1 How IC2 Outputs Operate

Clock Pulse	Q0 (Pin 3)	Q1 (Pin 2)	Q2 (Pin 4)	Q3 (Pin 7)	Q4 (Pin 10)
1	On	Off	Off	Off	Off
2	Off	On	Off	Off	Off
3	Off	Off	On	Off	Off
4	Off	Off	Off	On	Off
5	Off	Off	Off	Off	On
6	On	Off	Off	Off	Off
7	Off	On	Off	Off	Off

(IC3) to oscillate at a different frequencies, which creates sounds of different pitches. Because these outputs operate one at a time in sequence, you can create a sequence of four different sound levels to produce some interesting sound effects, as you will see shortly.

If you remove your finger from the switch (SW1), this takes resistor R2 out of the circuit and stops the 555 timer from producing pulses at pin 3, and this stops the sequencer from operating.

Notice that an additional capacitor C4 is included in this circuit, which is wired across the positive and negative battery supply. This capacitor is called a *decoupling capacitor*. The sound creation part of this circuit creates some electrical noise in the circuit, which could disrupt the operation of IC1 and IC2, which means that the circuit may not operate correctly. The inclusion of C4 helps to remove this electrical noise from the circuit and makes the circuit operation more reliable.

Things You'll Need

The components and equipment that you will need for this experiment are outlined in the following parts list. Prepare the items that you need before starting the experiment.

Code	Quantity	Description	Appendix Code
IC1 / IC3	1	555 timer IC	18
IC2	1	74HC4017 decade counter IC	20
R1	1	22KΩ 0.5W ±5% tolerance carbon film resistor	—
R2	1	220KΩ 0.5W ±5% tolerance carbon film resistor	—
R3–R6	4	Various 0.5W ±5% tolerance carbon film resistors with a resistance of more than 6.8KΩ	—
R7	1	6.8KΩ 0.5W ±5% tolerance carbon film resistors	—
R8	1	150KΩ 0.5W ±5% tolerance carbon film resistors	—
C1	1	Various capacitors with a capacitance of more than 10nF (minimum 10 volt rated)	—
C2	1	47nF ceramic disk capacitor (minimum 10 volt rated)	—
C3	1	100µF electrolytic capacitor (minimum 10 volt rated)	—
C4	1	220µF electrolytic capacitor (minimum 10 volt rated)	—
D1–D4	4	1N4148 signal diodes	—
LS1	1	Small loudspeaker, 66mm diameter (64Ω 0.3 watt rated)	22

(continued)

Things You'll Need (*continued*)

Code	Quantity	Description	Appendix Code
SW1	1	Normally open switch	24
B1	1	4.5V battery holder	15
B1	3	1.5V AA batteries	—
B1	1	PP3 battery clip	17
—	1	Breadboard	2
—	—	Wire links	—

NOTE

The Appendix Code column of the table refers to specific parts that I used in this experiment. Information about sourcing these parts is outlined in the Appendix.

The Breadboard Layout

NOTE

Refer to Chapter 3 for building breadboard layouts and fault-finding guidelines.

The breadboard layout for this experiment is shown in Figure 20-3. It includes the component codes to help you identify the components used in the circuit diagram and the parts list.

The close-ups in Figures 20-4, 20-5, and 20-6 show the breadboard layout from different angles, which should help you identify the pin connections.

FIGURE 20-3 The breadboard layout for the sound effects generator

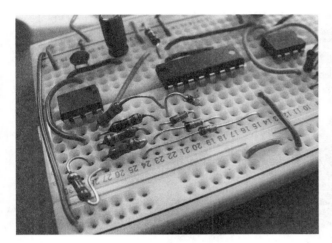

FIGURE 20-4 Close-up of the wiring around the sound effect diode and resistor network

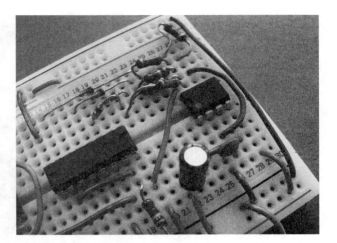

FIGURE 20-6 Close-up of the wiring around IC2 and IC3

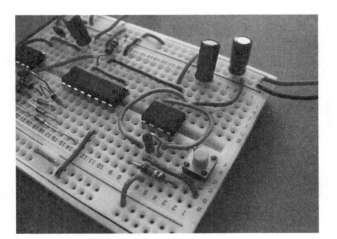

FIGURE 20-5 Close-up of the wiring around IC1 and SW1

When you are connecting the speaker to the breadboard, use the method that is shown in Chapter 10 to connect the speaker to the interconnecting wires.

The component values that you should use when you build this project are outlined in Table 20-2.

Figure 20-7 shows the final breadboard layout and also identifies where the three separate circuits (discussed earlier in this chapter) are located.

Time to Experiment!

Once you have built the breadboard layout, connect the battery to the circuit. You'll hear a tone coming out of the speaker. Now press the switch (SW1) and keep it pressed, and you should hear four different tones coming from the speaker, one at a time, a split second apart. This four-tone sound continues in a never-ending loop until you remove your finger and the noise stops on a single tone.

Now remove the battery and exchange the value of capacitor C1 for a 100nF ceramic disk capacitor. If you now reconnect the battery and press the switch, the sound sequence operates at a much faster rate and sounds a bit like a laser gun being fired or a red alert sound. How cool!

TABLE 20-2 Initial Component Values for R3, R4, R5, R6, and C1

Code	Quantity	Description
R3	1	82KΩ 0.5W ±5% tolerance carbon film resistor
R4	1	56KΩ 0.5W ±5% tolerance carbon film resistor
R5	1	150KΩ 0.5W ±5% tolerance carbon film resistor
R6	1	22KΩ 0.5W ±5% tolerance carbon film resistor
C1	1	100µF electrolytic capacitor (minimum 10 volt rated)

FIGURE 20-7 The location of the three separate circuit modules

Now remove resistor R3 from the circuit, so that only three tones are produced with a small gap in between, and press the button to see how this affects the sound. If you are familiar with the basic sound effects from some of the early arcade games such as Pac-Man or Space Invaders, you'll recognize these sound effects. Try removing one or more of the other resistors and see what effect this has.

Experiment with various capacitor values for C1 to see how they affect the sounds produced. Change the value of resistors R3 to R6 to see what different sound combinations you can create. Table 20-3 lets you note the various component values that you use for the five components so that you can use them again in the future.

TABLE 20-3 Record the Component Values for the Different Sounds That You Create

1 (R3)	2 (R4)	3 (R5)	4 (R6)	Speed (C1)	Sound Created
82KΩ	56KΩ	150KΩ	22KΩ	100nF	Space ship red alert sound

You should find that the lower the resistance value, the higher the pitch of the sound, and the higher the resistance value, the lower the pitch of the sound.

Further Experimentation

If you're feeling really inventive, you could use some of the other output pins of IC2 to create a ten-output sequencer rather than just four. To do this, you would need to remove pin 10 of IC2 from reset pin 15 and add six more signal diodes from each output before they are fed into each sound resistor. You could use this configuration to try and produce a short melody.

Summary

In this final experiment, you've learned how to combine a number of different circuit modules to create an interesting sound effects sequencer. If you have read this book from start to finish and you've completed all of the experiments, you should now be far more familiar with electronic components and circuits than you were at the beginning of the book. I hope that you enjoyed the experiments, and I wish you luck in your electronic endeavors. Happy inventing!

APPENDIX

Electronic Components and Suppliers

THE PARTS LISTED FOR EACH OF THE EXPERIMENTS in this book can be purchased from reputable electronic component suppliers. You will notice an Appendix Code column in the "Things You'll Need" table of each experiment; these refer to specific parts that I used in the experiments and are outlined in the table that follows. Basic components such as the various ½ watt rated carbon film resistors and the electrolytic and non-electrolytic capacitors don't show an appendix code, because these parts are readily available from many different electronics suppliers. It should be noted that whenever I used an electrolytic capacitor in the experiments in this book I used the *radial* type. The *radial* type of electrolytic capacitor

has its two leads protruding from one end of the component, unlike the *axial* type which has a lead protruding from either end of the device.

Because I am located in the United Kingdom, I acquired my parts from U.K. suppliers, several of which are listed in this section (along with one U.S. counterpart). To avoid high shipping costs, you'll likely want to get your parts from a supplier located within your country. I've listed the specific parts that I used simply as a reference to help you to identify and source them. For each part I've listed, you can search by part number on the respective supplier's web site to get a visual and technical details.

Breadboards

Code	Description and Manufacturer's Part Number	Supplier and Part Number
1	Breadboard (AD-100)	Maplin Electronics AD-100
2	Breadboard (AD102)	Maplin Electronics AD102 (AG10L)
3	Breadboard (AD-01)	Maplin Electronics AD-01 (BZ13P)

LEDs

Code	Description and Manufacturer's Part Number	Supplier and Part Number
4	5mm red LED	RS Components 228-5972
5	5mm green LED	RS Components 228-6004
6	5mm RGB LED	RS Components 247-1511
7	5mm high-intensity white LED, 6000mcd	RS Components 668-6338
8	Seven-segment red common cathode display (SC05-11EWA)	RS Components 235-8783

Transistors

Code	Description and Manufacturer's Part Number	Supplier and Part Number
9	BC108C NPN transistor	ESR Electronic Components BC108C
10	BC178 PNP transistor	ESR Electronic Components BC178

Sensors

Code	Description and Manufacturer's Part Number	Supplier and Part Number
11	5KΩ @25°C NTC 500mW disc thermistor	ESR Electronic Components 928-250 RS Components 706-2787
12	Light-dependent resistor (NORPS-12)	RS Components 651-507
13	Infrared photodiode and amplifier (PNA4602M)	RS Components 199-630

Battery Holders

Code	Description and Manufacturer's Part Number	Supplier and Part Number
14	AA battery holder (two AA batteries, 3 volts)	RS Components 512-3580
15	AA battery holder (three AA batteries, 4.5 volts)	Maplin Electronics YR61R
16	AA battery holder (four AA batteries, 6 volts)	RS Components 594-432
17	PP3 battery clip	RS Components 489-021 (pack of 5)

Integrated Circuits

Code	Description and Manufacturer's Part Number	Supplier and Part Number
18	555 timer	RS Components 534-3469
19	4026B decade counter with seven-segment display outputs	ESR Electronic Components 4026B
20	74HC4017 decade counter	RS Components 709-3062 (pack of 10)

Earpieces and Speakers

Code	Description and Manufacturer's Part Number	Supplier and Part Number
21	3MΩ crystal earpiece	ESR Electronic Components 203-004 Maplin Electronics LB25
22	Small loudspeaker, 66mm diameter (64Ω 0.3 watt rated)	ESR Electronic Components 203-001

Switches and Relays

Code	Description and Manufacturer's Part Number	Supplier and Part Number
23	6 volt DC coil, DPDT 1A 24v DC subminiature DIL relay (SRC-S-06VDC)	ESR Electronic Components 242-340
24	6mm-by-6mm momentary push to make normally open switch (PCB mounted)	RS Components 479-1441 (pack of 20) ESR Electronic Components 212-050

Sounder

Code	Description and Manufacturer's Part Number	Supplier and Part Number
25	6 volt 25mA 75dB reed buzzer	ESR Electronic Components 030-012

Miscellaneous

Code	Description and Manufacturer's Part Number	Supplier and Part Number
26	Double-sided adhesive sponge strips	RS Components 512-856 (bag of 25)

Suppliers

Here are the contact details of the various suppliers that I used to acquire my parts:

RS Components Ltd.
www.rswww.com

Allied Electronics, Inc.
(U.S. distributor owned by same group that owns RS Components)
www.alliedelec.com

(Note that RS Components part numbers may not be the same as Allied Electronics part numbers.)

ESR Electronic Components Ltd.
www.esr.co.uk

Maplin Electronics Ltd.
www.maplin.co.uk

Index

Tomorrow's scientists and engineers play with Thames & Kosmos kits today!

thamesandkosmos.co

THAMES & KOSMOS